NATIONAL DEFENSE RESEARCH INSTITUTE

T0308377

Maritime Tactical Command and Control Analysis of Alternatives

Bradley Wilson, Isaac R. Porche III, Mel Eisman, Michael Nixon, Shane Tierney, John M. Yurchak, Kim Kendall, James Dryden, Sean Critelli

Prepared for the United States Navy

Approved for public release; distribution unlimited

For more information on this publication, visit www.rand.org/t/RR1383

Library of Congress Cataloging-in-Publication Data is available for this publication.
ISBN: 978-0-8330-9572-5

Published by the RAND Corporation, Santa Monica, Calif.
© Copyright 2016 RAND Corporation
RAND® is a registered trademark.

Cover: *Officers aboard the guided-missile destroyer USS Ross (U.S. Navy photo by Mass Communication Specialist 2nd Class John Herman/Released).*

www.rand.org

Preface

This report is about software systems that support situational awareness and command and control (C2) afloat and ashore. Specifically, this report provides the results of an analysis of alternatives for the follow-on to the U.S. Navy's Global C2 system. The replacement is intended to be procured as an ACAT III program of record called *Maritime Tactical Command and Control* (MTC2).

This research was sponsored by the Department of the Navy, and conducted within the Acquisition and Technology Policy Center of the RAND National Defense Research Institute, a federally funded research and development center sponsored by the Office of the Secretary of Defense, the Joint Staff, the Unified Combatant Commands, the Navy, the Marine Corps, the defense agencies, and the defense Intelligence Community.

For more information on the RAND Acquisition and Technology Policy Center, see www.rand.org/nsrd/ndri/centers/atp or contact the director (contact information is provided on the web page).

Contents

Figures

Tables

Summary

Problem Statement

Global Command and Control System-Maritime (GCCS-M) is the U.S. Navy's legacy command and control (C2) system to provide situational awareness (SA) and tracking of blue forces. It includes both hardware and software. GCCS-M, the maritime implementation of U.S. GCCS, is focused primarily on the SA capability in the form of the common operational picture (COP), for example, "tracks on a map." It provides commanders at all echelons a single, integrated, scalable command, control, communications, computers, and intelligence (C4I) system. The COP is shared across more than 75 command, control, communications, computers, cyber, intelligence, surveillance, and reconnaissance systems ("PMW 150 Command and Control Systems Program Office," 2015).

Legacy C2 systems are not sufficient enablers of projected maritime C2 capability needs. Like the entire GCCS family of systems, GCCS-M is perceived as "not likely to satisfy Joint Command and Control (JC2) capability needs as articulated in the JC2 Capability Development Document" (Walsh et al., 2005).

As the Navy migrates to a common hardware platform,[1] GCCS-M must modernize to fit the new software-only paradigm. The Maritime Tactical Command and Control (MTC2) program is being pursued to provide a more-robust set of C2 capabilities using the common hardware footprint and web-based interfaces. The new program intends to

---

[1] Via the Consolidated Afloat Networks and Enterprise Services (CANES) program of record.

phase out the Navy's implementation of GCCS-M. The focus is on rapid fielding at lower cost using a software-only program. Additionally, MTC2 intends to bring a wider range of C2 functions beyond SA to include readiness, intelligence, planning, and tasking through an interactive "halo" COP. The halo COP is an expanded COP (or E-COP) that includes mission-specific readiness capabilities, networks, and intelligence for the items being tracked (Akins, 2011).

Objective

The Navy articulated four alternatives as a set of reasonable options to the current legacy system ("OPNAV Study Guidance," 2013). The alternatives considered include:

1. Maintain the legacy system in the fleet today.
2. Leverage developments by the Office of Naval Research (ONR) called *Command and Control Rapid Prototyping Continuum* (C2RPC) and augment the current legacy system.
3. Create a new development program that expands the capabilities of the legacy system with the robustness of new designs via a linkage to an ashore cloud.
4. Create a new development program that expands the capabilities of the legacy system but with the robustness of new designs that will enable a linkage to both an afloat and an ashore cloud. The afloat cloud is called the *Naval Tactical Cloud (NTC)*.

These alternatives are evaluated based on their life-cycle cost, risk, information assurance (IA) characteristics, and their ability to meet the performance outlined in the initial MTC2 guidance.

Approach

Modeling and simulation within the context of a major combat operation was performed on the alternatives using a discrete event simulator—

called the *Process and Architecture Analysis Tool (PAAT)*—embedded with an Office of the Chief of Naval Operations N81–approved scenario. The three key Navy work processes used to explore the alternatives within the context of the scenario were

- provide commander's update
- develop a personnel recovery mission
- update and/or manage COP tracks.

These work processes were selected for a number of reasons. As MTC2 is a C2 system, it was important to capture aspects of the commander's decision cycle. *Joint Publication 3-33* describes this cycle in its simplest form as monitor, assess, plan, direct (Joint Chiefs of Staff, 2012). Each work process touches on aspects of the decision cycle, some more so than others. Additionally, the selection of work processes had constraints of time and availability of data. Planning processes were not always well documented and can vary depending on the mission, the commander, and the Maritime Operations Center (MOC).

Performance results were calculated using measures of effectiveness derived from mission tasks (MTs) drawn from initial MTC2 guidance, which sourced the *Command and Control, Joint Integrating Concept* (U.S. Department of Defense [DoD], 2005). The MTs and measures are detailed in Table S.1.

The MTs and the corresponding measures provided quantitative comparisons of the alternatives. In addition, tabletop exercises, risk assessments, and IA assessments were conducted using subject-matter experts and reviewing existing design and design options. The risk assessment judged each alternative on a number of risk areas and scored them as low, medium, or high risk, according to U.S. Air Force *Analysis of Alternatives (AoA) Handbook* recommendations (Office of Aerospace Studies, 2010). The risk assessment is broken into multiple subcategories according to handbook guidance that notes "the qualitative evaluation of risk consequence will be determined by the criteria shown" (Office of Aerospace Studies, 2010, p. 40). The subcategories were assessed qualitatively by analysts at the RAND Corporation using

Table S.1
Mission Tasks and Measures

Mission Tasks	Measures
1. Plan collaboratively	1-1: Percentage of data and information that is interoperable between joint, interagency, intergovernmental, and multinational (JIIM) partners
	1-2: Timeliness of planning information dissemination to JIIM partners
2. Develop and maintain shared SA and understanding	2-1: Percent of friendly force locations that are accurate; timeliness as a proxy for accuracy
	2-2: Completeness of COP (percentage of forces)
	2-3: Percentage of available bandwidth consumed
3. Establish/adapt command structures and enable both global and regional collaboration	3-1: Percentage of C2 data that is interoperable across tactical/operational echelons
	3-2: Likelihood of being able to recon to dynamic mission requirements and returning to steady state
4. Communicate commander's intent and guidance	4-1: Percentage of time orders are received in time to conduct the task/mission
	4-2: Percentage of commander's plans received by maritime commanders
5. Exercise command leadership	5-1: Time to promulgate rules of engagement and rules for the use of force changes
	5-2: Percentage of released information that is correct (accuracy)
	5-3: Number of readiness assessments completed in time

Table S.1—Continued

Mission Tasks	Measures
6. Synchronize execution across all domains	6-1: Percentage of subordinate forces able to access unclassified information at operational level (accessibility)
	6-2: Percentage of subordinate forces able to access unclassified information at the tactical level (accessibility): 95%
	6-3: Plans are completed, disseminated, received in time
7. Monitor execution, assess effects, and adapt operations	7-1: Percentage of forces and assets that can quickly change operations to facilitate direction change (agility)
	7-2: Number of fires processes, networks, and systems that Maritime MOC can efficiently track
8. Leverage mission partners	8-1: Percentage of mission partners that receive and understand commanders' intent
	8-2: Percentage of JIIM partners MOC can exchange information with

SOURCE: Initial MOC2 guidance and DoD, 2005 (shown are the metrics that align with approved analysis of alternatives (AoA) measures of effectiveness (MoEs) and measures of performance (MoPs).

the criteria, and scores for overall technical, schedule, programmatic, and cost risk were developed.

Finally, a cost assessment was completed using data provided by PEO C4I Program Management, Warfare (PMW) 150. The cost analysis is primarily based on software sizing and reuse factors from an estimate of source lines of code provided to the study team. The estimate covers the PMW 150 software-only development cost. Installations begin in fiscal year (FY) PMW 2014 at shore and afloat sites as designated in the MTC2 fielding inventory objective plan. The cost estimate considered software growth estimates, as well as MTC2 top-level system specifications, and software interface requirements for the CANES stack. In Alternatives 1 and 2, GCCS-M Increment 2 sustainment cost is considered through FY 2030. In Alternatives 3 and 4, the cloud-based environments are assumed to be paid for by the PMW 120. Additionally, sustainment for GCCS-M begins to phase out starting in FY 2018 in Alternatives 3 and 4.

Assessment Summary of Alternatives

Alternative 1
The legacy system performed poorly compared with the other alternatives. It had no new capabilities and no enhancements to its data architecture. Without the benefit of a comprehensive data strategy, it could not meet future demands for information interoperability. Furthermore, without new capabilities to automate the processing and presentation of information—to facilitate efficient workflows that would further enhance the productivity of operational specialists and commanders and others who rely on such C2 systems—it could not meet timeliness demands for collaborative planning, establishing and adapting command structures, and communicating the commander's intent. It performed comparably well to the other alternatives in providing friendly force SA, a task that it does relatively well today. Alternative 1 had no development costs, but annual sustainment costs after MTC2 reaches full deployment were five times higher than projections for Alternatives 3 and 4. Alternative 1's risks are considered highly prob-

able, given that only limited modernization and investment will take place. Alternative 1's risks identify many areas where the legacy system will be unable to meet future capability requirements, for example, operating in a denied communication environment, implementing multilevel security, and providing collaborative planning tools.

Alternative 2

Relative to the baseline (Alternative 1), Alternative 2 performed better in terms of key MTs such as establish and adapt command structures to enable global and regional collaboration because it did bring some additional capabilities with C2RPC. However, many of these capabilities did not perform robustly enough to meet the effectiveness objectives because, like Alternative 1, they did not benefit from workflow efficiencies and information interoperability brought on by adhering to an enterprise-wide information-sharing architecture. Alternative 2 had the highest overall life-cycle costs due to maintaining GCCS-M and augmenting it with C2RPC. By maintaining the legacy system, Alternative 2 has high-probability risks of not implementing a multilevel security solution, employing automated methods of handling information, and scaling to support increasing numbers of data feeds.

Alternative 3

Alternative 3 benefited from the enhanced interoperability that an ashore cloud can provide. This contributed to workflow efficiencies that enhanced productivity of operational specialists and commanders alike, specifically in the area of collaborative planning. The weakness of Alternative 3 is that the bulk of the data is stored ashore, potentially making planning at sea cumbersome. This is observable in Table S.1, MoE 1–2, where Alternative 3's performance-disseminating planning products to mission partners lagged behind Alternative 4. Efficient methods for transferring and synchronizing information, particularly on low-bandwidth and/or high-latency links, would need to be established. As a new system for replacing GCCS-M, this alternative results in potential annual sustainment cost savings through FY 2030. Additionally, both Alternatives 3 and 4 leverage a Navy cloud effort that is funded from outside the MTC2 program. This provides substantial cost savings. Alternative 3's risks are of higher consequence but lower probability than those found in Alternatives 1 and 2. These challenges include development of a comprehensive data strategy; a capable data integration layer; and functioning in a disconnected, intermittent, or limited communications environment. A sample of IA assessments acquired through in-person interviews suggests that sustainment of Alternative 3 will be more feasible than Alternatives 1 and 2, but a more thorough IA analysis is needed.

Alternative 4

Alternative 4 benefited from the enhanced interoperability that an ashore and afloat cloud can provide. This contributed to workflow efficiencies that enhanced productivity of operational specialists and commanders alike. In some areas this alternative performed considerably better than Alternative 3, as well as the legacy alternatives. Alternative 4 provides pervasive access to information and analytics in a common, globally managed enterprise, and demonstrates the greatest potential for achieving C2 workflow efficiencies. The tabletop results agreed with findings in modeling and simulation and in interoperability analysis with regard to identifying the advantages of Alternative 4.[2] The likelihood of being able to reconfigure to dynamic mission requirements and return to steady state was most probable in Alternative 4. The increased probability is due to expanded planning functionality at MOC and afloat locations, as well as infrastructure that better enables inter-echelon C2 information movement. Alternative 4 also provides the greatest potential to operate in disconnected, interrupted, and low-bandwidth (DIL) environments.

Alternative 4 was less expensive than Alternatives 1 and 2 with regard to annual sustainment cost savings through FY 2030. In regard to risk, Alternative 4 has similar problems as Alternative 3: Because of the inclusion of a dependency on NTC, delays in that circumstance may be outside the control of the program, hence the schedule and programmatic risks. Historically, the Navy has managed schedule and program risk in some ways by stovepiping processes within a program.

---

[2]   See the Appendix for more information on these results.

This can have the adverse consequence of increasing costs and decreasing interoperability. By forcing application providers to use commodity hardware such as what CANES provides, the Navy is increasing dependencies in the hopes of reducing cost and increasing interoperability. Managing the programmatic and schedule risk of dependency on CANES should be similar in construct to a dependency on Navy cloud efforts. As this analysis has shown, if the Navy can manage risk effectively, performance gains are possible.

A sample of IA assessments acquired through in-person interviews suggests that sustainment of Alternative 4 will be more feasible than Alternatives 1 and 2, but a more-thorough IA analysis is needed.

Performance Measures

Overall alternative performance is shown in Figure S.1.

In the modeling and simulation of the personnel recovery scenario, planning information was disseminated to JIIM partners four times faster in Alternative 4 over Alternative 3. Also, orders were received in time to conduct the mission significantly more often in Alternative 4 than Alternative 3, and overwhelmingly more often than Alternatives 2 and 1.

The commander's update brief preparation and planning information was disseminated to JIIM partners one and a half times faster in Alternative 4 over Alternative 3.

In regard to accuracy, completeness and network use, the differences in the performance between the alternatives for management of the COP were not significant. The ability to meet the requirements for accuracy and completeness for the phase of the scenario simulated was generally good.

The results from the performance,[3] cost, and risk analysis are a preference for Alternatives 3 and 4. The expectation is that MTC would not bear the cost burden of the cloud, and there were no common or unique software tools or applications specifically identified for shore- or tactical-only clouds, hence the lack of demarcation between the two alternatives on cost. We discuss these limitations further and areas for

---

[3] Including modeling and simulation, static analysis, and the tabletop exercise.

Figure S.1
Analysis of Alternative Summary Results

	Critical measures									PMW 150 life-cycle cost estimate (FY 2013 $) FYs 2014–2030 (confidence level 50%)	Risk				IA considerations				
	MT1 Plan collaboratively		MT2 Develop and maintain shared SA and understanding			MT3 Establish/adapt command structures and enable both global and regional collaboration		MT4 Communicate commander's intent and guidance											
	1-1	1-2	2-1	2-2	2-3	3-1	3-2	4-1	4-2											
	Interoperability	Timeliness	Accuracy	Completeness	Bandwidth efficiency	Interoperability	Responsiveness	Speed	Timeliness		Technical	Schedule	Programmatic	Cost	Architecture finalization date	Inherited IA protections	Security posture maturity	Security posture sustainability	Level of app/infra coupling
Alt 1: Status quo. GCCS-M increment 2 modernization will be limited.										$407									
Alt 2ᵃ: Augment GCCS-M increment 2 capabilities with productized C2RPC capabilities and newly developed capabilities following C2RPC style and approach.										$464									
Alt 3: New system to satisfy maritime C2 requirements as defined in the initial MTC2 guidance. Builds upon the IC data analytic cloud capability. This system will replace GCCS-M at every program of record site.										$369									
Alt 4: New system to satisfy maritime C2 requirements as defined in the initial MTC2 guidance. Adds the concept of ONR Tactical Cloud.										$359									

NOTES: See Table 3.3 for information about color coding for the critical performance measures. See Table 4.11 for information about the color coding for risk. See Figure 5.10 for color coding for IA considerations.

ᵃ For report, Alternative 2 assumes newly developed capabilities, "2+."

refinement in Chapter Four. There is a preference for Alternative 4 for in risk and performance. It provides pervasive access to information and analytics in a common, globally managed enterprise, and demonstrates the greatest potential for achieving C2 workflow efficiencies.

Acknowledgments

The authors of this report would like to thank Space and Naval Warfare Systems Command (SPAWAR) study lead Robert Parker, assistant program executive officer, science and technology (APEO S&T) and study director of Maritime Tactical Command and Control (MTC2). Thanks also go to Commander, U.S. Naval Forces Europe-Africa/U.S. 6th Fleet (CNE-CNA-C6F) for hosting information-gathering sessions. Specifically, we are grateful to their science advisor, Gary R. Shaffer, for hosting, organizing, and contributing in discussions with a range of staff. We also would like to thank Andrew Pruiett, knowledge management officer, CNE-CNA-C6F, for providing workflow documentation, as well as valuable extended discussion with the team.

Maritime Operations Center (MOC) 3rd Fleet personnel allowed us to shadow their activities and shared data and insight with our study team early in our research. Joseph Spencer, Tracy Conroy, LCDR Mark Kaul, and LCDR Nathan Walker enabled that interaction.

In addition, we thank others at SPAWAR who engaged us individually to provide insight, suggestions, and direct assistance, including Gleason Snashall (SPAWARSYSCEN), Dave Gorman, Phil Summerly, Floyd Pinkney (PEO C4I), Timothy Jara (SPAWARSYSCEN), Joaquin Martinez de Pinillos, Josh Gomer, and Mike Davis (SPAWAR).

At Naval Engineering Logistics Office (NELO), Sally Mahdavi and Michella Hensley shared their thoughts on cost issues. Finally at the RAND Corporation, Michelle McMullen, Paul DeLuca, and Cynthia Cook helped edit and revise early drafts, and Amado Cordova and Christopher Mouton provided thoughtful reviews.

Abbreviations

ACS	Agile Core Services
A2AD	anti-access and area of denial
AoA	analysis of alternatives
C2	command and control
C2RPC	Command and Control Rapid Prototyping Continuum
C4I	command, control, communications, computers, and intelligence
CANES	Consolidated Afloat Networks and Enterprise Services
CARD	Cost Analysis Requirements Document
CDD	Capability Development Document
CoA	course of action
COP	common operational picture
COTS	commercial off-the-shelf
DCGS-A	Distributed Common Ground System-Army
DCGS-N	Distributed Common Ground System-Navy
DIL	disconnected, interrupted, and low bandwidth
DoD	U.S. Department of Defense

E-COP	expanded common operational picture
ELT	Extract, Load, Transform
FDD	final delivery date
FY	fiscal year
GCCS-M	Global Command and Control System-Maritime
GOTS	government off-the-shelf
IA	information assurance
ICW	Interactive Courseware
JC2	Joint Command and Control
JCA	Joint Capability Area
JIIM	joint, interagency, intergovernmental, and multinational
LCC	life-cycle cost
MANA	Map Aware Non-Uniform Automata
MDA	milestone decision authority
MOC	Maritime Operations Center
MoE	measure of effectiveness
MoP	measure of performance
MSFD	Multi Service Force Deployment
MT	mission task
MTC2	Maritime Tactical Command and Control
NCC	Navy Component Commander
NTC	Naval Tactical Cloud
ONR	Office of Naval Research

OPLAN	operation plan
OTM	Open Track Manager
PAAT	Process and Architecture Analysis Tool
PEO C4I	Program Executive Office for Command, Control, Communications, Computers, and Intelligence
PLCCE	program life-cycle cost estimate
PMW	Program Management, Warfare
RoE	rules of engagement
S/W	software
SA	situational awareness
SLOC	source lines of code
SME	subject-matter expert
SPAWAR	Space and Naval Warfare Systems Command
SSA	Software Support Activity

CHAPTER ONE

Introduction

1.1. Background

Command and control (C2) is the means by which a commander synchronizes and/or integrates force activities in order to achieve unity of effort. Maritime commanders from the U.S. Navy Component Commander (NCC) down to the tactical unit or element commander must exercise C2 of assigned Navy, joint, and coalition forces.[1]

Legacy C2 systems are not sufficient enablers of projected maritime C2 capability needs. Global Command and Control System-Maritime (GCCS-M) is the current maritime C2 system providing C2 from the operational level to the tactical edge. Like the entire GCCS family of systems, it is perceived as "not likely to satisfy JC2 capability needs as articulated in the JC2 CDD" (Walsh et al., 2005).

Furthermore, the architecture has been described as requiring a large number of individual interfaces that are difficult to change, and a process for maintaining situational awareness (SA) that is labor intensive. This analysis came out of the Joint Command and Control Analysis of Alternatives (AoA), an effort that started in the early 2000s to manage similar capabilities across the DoD enterprise, where one of the capabilities was joint C2, and the recommendation was that a JC2

---

[1] Joint and Navy doctrine both define C2 to cover a broad range of requirements, which are defined in Joint Staff, *Joint Command and Control (C2) Requirements Management Process and Procedures*, Washington, D.C., CJCSM 3265.01A, November 29, 2013, and U.S. Department of Defense (DoD), *Command and Control, Joint Integrating Concept*, Final Version 1.0, September 1, 2005.

program should be pursued. The Joint Command and Control Capability AoA was conducted in 2004. The JC2 AoA identifies joint command and control (C2) capability gaps by comparing joint C2 capability needs, as articulated in the Joint Command and Control (JC2) Capability Development Document (CDD), with the existing capabilities of the Global Command and Control System Family of Systems (GCCS FoS). The objective was that "GCCS will evolve from its current state of joint and Service variants to a single Joint C2 architecture and capabilities-based implementation" (Wellman, 2005).

Eventually, the JC2 program emerged as the Net-Enabled Command Capability, a joint program envisioned to be the replacement for GCCS, but that program suffered numerous problems and was cancelled because

> it was at significant risk of not being able to deliver capabilities to meet validated warfighter requirements and was not able to meet its Initial Operational Capability within schedule. Instead, the DoD will focus the Joint Command and Control (C2) efforts on consolidating the systems and technologies of the NECC program into the Global Command and Control System (GCCS) family of systems. The approach will be an incremental, spiral approach to modernizing the GCCS family of systems, deploying modular, operationally useful, and tested capabilities while moving towards a net-centric, web-based, standards-based service oriented architecture. The funding is being redirected to support sustainment of the current Global Command and Control System – Joint (GCCS-J) Family of Systems (FoS) to ensure the sustainment and synchronization of activities required to maintain a robust command and control program. (Defense Information Systems Agency [DISA], 2010)

As a result, another AoA, called the Joint C2 AoA, was conducted and completed in 2011. Joint C2 AoA relies on a federation of systems, hence the Navy's planning for Maritime Tactical Command and Control (MTC2) as their service's follow-on to GCCS-M. The analysis in both AoAs is useful and applicable to the various GCCS variants in existence today, notably GCCS-Joint, of which the GCCS-M code base is built upon.

GCCS-M, the maritime implementation of U.S. GCCS, is focused primarily on the SA capability in the form of the common operational picture (COP), for example, "tracks on a map." It provides commanders at all echelons a single, integrated, scalable command, control, communications, computers, and intelligence (C4I) system. The COP is shared across more than 75 command, control, communications, computers, cyber, intelligence, surveillance, and reconnaissance systems ("PMW 150 Command and Control Systems Program Office," 2015).

MTC2 is expected to be an ACAT III program of record within the Program Executive Office (PEO) for C4I. The new program intends to phase out the Navy's implementation of GCCS-M. The focus is on rapid fielding at lower cost using a software-only program. The Navy intends to supply software applications with commoditized hardware through the Consolidated Afloat Networks and Enterprise Services (CANES) program, and legacy hardware and software systems, such as GCCS-M, must adapt to this new architectural paradigm.[2] Historically, many application providers brought their own hardware to afloat platforms, creating stovepipes that limited reuse of code and came with additional cost. By leveraging CANES, the Navy intends to make use of the increase in computing power and decrease in cost that has driven the commercial industry for decades.

Additionally, MTC2 intends to bring a wider range of C2 functions beyond SA to include readiness, intelligence, planning, and tasking through an interactive "halo" COP.

The halo COP is an expanded COP (or E-COP) that includes mission-specific readiness capabilities, networks, and intelligence for the items being tracked (Akins, 2011). An example is illustrated in Figure 1.1.

---

[2]   CANES is the Navy's next-generation tactical afloat network, consolidating five legacy networks into one. It includes a common computing environment, which applications such as MTC2 will run on, as well as common software services that application program providers can leverage.

The halo is an interactive icon on a user interface overlaid on a map of the world. Users can interrogate icons on the map to activate the halo and explore additional data about the entity. This may include network operations, mission readiness, tasks being performed, people, supplies, etc. The intent is to bring an additional layer of data to users and ultimately decisionmakers to increase their SA.

The MTC2 program goal is to provide an evolutionary C2 solution that seeks to be more responsive to fleet needs, support rapid software releases, and enable agile response to requirements modifications.

One major enhancement for the MTC2 program to consider is to incorporate prototype capabilities from the C2 Rapid Prototyping Continuum (C2RPC), which is an Office of Naval Research (ONR) development. C2RPC is a product and an approach; the latter has been described as a "new method for developing and implementing Command and Control (C2) software." It relies on an active partnership among Commander, U.S. Pacific Fleet, ONR, and the C2 Program Office in the Program Management, Warfare (PMW) 150 for PEOC4I. According to the stakeholders involved, the partnership serves as an incubator for technology concepts that is rapidly producing capabilities to be transitioned to C2 programs of record. It has already resulted in the development of prototype C2 applications such as Open Track Manager (OTM), E-COP, and Plans and Task Data Services. The C2RPC team brought together full-time subject-matter experts (SME) in SA, Navy planning and assessment, fleet readiness, and Maritime Operations Center (MOC) operations (ONR, 2012). The entire suite of tools coming out of C2RPC is sometimes also referred to as C2RPC, meaning a product as well as an approach.

C2RPC is using a distributed, service-oriented architecture approach to provide basic services needed for C2. This contrasts with the architecture of GCCS-M, which is a more monolithic pattern. Updating the GCCS-M track management system requires an update to the entire package of GCCS-M software. Whereas an update to the OTM service can be done more cleanly in isolation to other services because its interfaces are explicitly defined. This makes the overall architecture of a system implementing the various services coming out of C2RPC more modular. The OTM "provides enhanced track

Figure 1.1
Examples of "Halos"

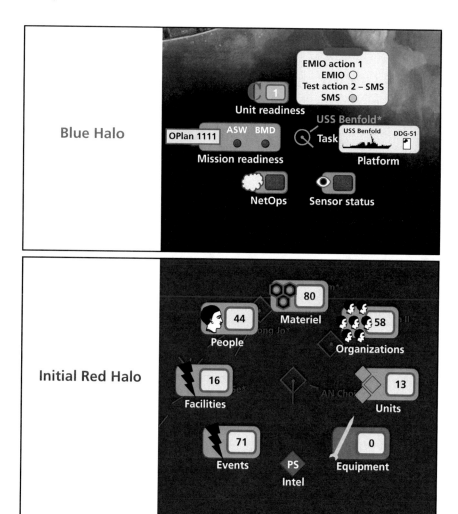

Blue Halo	
Initial Red Halo	

management performance, functionality and connectivity to a variety of legacy and future COP." The interrelated E-COP "allows the C2RPC capabilities to communicate through the shared plan representation and user-facing services (Halo COP) of the C2RPC architecture" (ONR, 2012). Finally, the plan and task data services are "a set of viewers and editors that links planning and execution information to the Halo COP so that planners can access frequently asked questions information from the Halo COP and enter important information which links various data sources into a common representation onto other data views" (ONR, 2012). The latter piece regarding planning data and mission-oriented tasks is something that is new in the C2 tool and unavailable in GCCS-M. It exposes users to the potential space of the mission-planning process and many of the artifacts that are potentially available, such as current and future operations plans, commander's update briefs, etc. This capability was discussed in the JC2 AoA abstract as a need for "deployment planning capabilities."

1.2. Purpose

This report explores and analyzes alternative approaches to achieving future Navy objectives and missions supported by MTC2. The Navy articulated the four alternatives, but it was left to the research team to add, remove, or modify them as necessary ("OPNAV Study Guidance," 2013). The study analyzes and scores the alternatives based on performance, cost, risk, and information assurance (IA) considerations.

1.3. Scope

In accordance with the guidance for this AoA, the analysis will explore solutions to meet maritime C2 capability gaps and/or deferred attributes in the following documents:

- JC2 Capability Development Document
- *Command and Control, Joint Integrating Concept* (DoD, 2005)
- Initial MTC2 Guidance.

Initial MTC2 guidance extends the capabilities of the GCCS-M Increment 2 by adding enhanced SA, readiness, planning, monitoring, and assessment capabilities.

The AoA base lined current GCCS-M 4.1 capabilities; assessed the cost effectiveness of reasonable alternatives; conducted full consideration of possible trade-offs among cost, schedule, and performance objectives for each alternative; and identified areas of technical risks that are candidates for prototyping. In areas where the Navy has already commissioned related analysis efforts, the AoA incorporated and leveraged results as appropriate. Those results are described in Chapter Four.

1.4. Study Team/Organization

The RAND National Defense Research Institute conducted the MTC2 AoA by developing an analytical comparison of the operational effectiveness, cost, and risks of four specific alternatives. Figure 1.2 depicts the basic organizational structure of the MTC2 AoA. We liaised with stakeholders as required, obtaining the needed subject matter expertise.

Office of the Chief of Naval Operations Deputy Chief of Naval Operations for Information Dominance (N2/N6), as the sponsor of the MTC2 AoA:

- developed the MTC2 AoA study plan, which set forth the alternatives under study
- had an N2/N6 advisory representative on the study team.

PEO C4I, as the milestone decision authority (MDA):

- approved the AoA Study Plan in coordination with the Deputy Director, Assessment Division (N81B)
- established the MTC2 AoA team
- co-approved the AoA final report in coordination with N81B

- (N81 was offered an advisory role on the study team).

The AoA study director:

- acted as the spokesperson for the AoA by presenting status brief-ings
- organized the study team
- acted as the single authority for providing direction to the study team
- oversaw the preparation and accuracy of the data included in the AoA report
- presented the results of the analysis to senior Navy leadership.

1.5. AoA Review Process

The AoA review process involved three interim progress reports over the course of the 180-day study. A final briefing to PMW 150 leadership enabled a review of cost, risk, and performance assessments. Subsequent revisions were put in place, and a final briefing was given to the MDA on June 20, 2013.

Figure 1.2
MTC2 AoA Organizational Structure

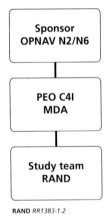

RAND *RR1383-1.2*

1.6. Scenarios

A 2016 Multi Service Force Deployment (MSFD) scenario was employed to provide force laydowns and context within the simulation and modeling. It was also used as background for the static analysis and tabletop gaming. Force complements and positions were used to inform the blue, red, white, and green COPs. COP data, along with ISR information and background traffic, drive more-realistic network traffic and manpower use that affects the time line of mission tasks (MTs).

As described in the February 2013 study plan, the scenario exercises the following capabilities:

1. Continuous maintenance and display of the COP for the AOR. The COP includes maritime and air, friendly, coalition, enemy, unknown tracks, and commercial vessels of interest.

2. Continuous planning, execution, and monitoring. This begins with the analysis of the strategic directive received from headquarters, mission readiness assessment of the forces, and formulation and dissemination of the Commander's Initial Guidance and Intent. It continues with the development of courses of action (CoAs), selection of the CoA, collaborative planning, and dissemination of plans and updates, along with the Commander's Intent. Execution is continuously monitored and evaluated, and plans are updated and disseminated.

3. Knowledge management that ensures efficient, timely, and comprehensive knowledge is shared within the capacity and latency limits of the networks. Shared information includes, but is not limited to, the Commander's Intent and Guidance, rules of engagement (RoE), political and cultural information.

4. The use and exploitation of the data cloud. It is expected that other systems will publish relevant data to the cloud (e.g., intelligence products, readiness information, RoEs, historical data, etc.).

The following ground rules, constraints, and assumptions were provided for the AoA.

1. The AoA study team will leverage findings from past studies and ongoing analysis efforts to inform the AoA as appropriate.
2. This effort must reference all previous analyses used to support the preferred strategy. All source documents must be cited and delivered with the final AoA report.
3. For those alternatives recommended for elimination, the analysis will only be completed to the point of demonstrating the rationale for exclusion, with the exception of Alternative 1 (status quo), which must be completed to establish the baseline for future comparison. The basis for exclusion must be fully justified in the AoA report.
4. As a software-only program, it is anticipated that the preferred alternative for the maritime C2 system will rely on hardware provided by CANES on afloat platforms and CANES and/or MOC Enterprise LAN Solution (MELS) at shore sites.
5. The specific cost-estimating methodology for each alternative may vary based on the available data but can include engineering buildup, parametric estimating techniques, and the analogy technique. The cost estimates must be sufficiently detailed to rank the alternatives but are not expected to be of budget quality.
6. All data will be normalized to establish the basis for all subsequent analysis, model building, and creation of decision-support tools. Data will be normalized in the most-appropriate manner, including adjusting for inflation, quantity, assessing schedule differences, and customizing for technical characteristics where appropriate.
7. The analysis will include the training requirements needed to implement each alternative.
8. The analysis should include an evaluation of system security to include but not limited to cyber protection, software vulnerabilities, and program protection. Additionally, the alternatives should address system reliability of each alternative.

9. The analysis should include an evaluation of compliance with the Joint Objective Architecture, CANES, and the DoD Chief Information Officer Information Technology Consolidation plan, and leveraging of common services and capabilities provided.

10. The study team will take direction only from the study director and shall inform the study director and PMW 150 in writing of any direction they have received that will impact their ability to complete the effort in accordance with the study schedule.

CHAPTER TWO

Alternatives

2.1. Description of Alternatives

The alternatives were provided in the study guidance by OPNAV N2N6F4 ("OPNAV Study Guidance," 2013). They are defined as follows:

1. The status quo alternative shall serve as the baseline for comparing and evaluating all the alternatives. The basis for the status quo shall be GCCS-M Increment 2 projected costs and capabilities already fielded as well as any enhancements already programmed. The primary function of GCCS-M Increment 2 is to provide maritime SA augmented with a small number of basic tactical decisions aids. There will be no more follow-on capability developments and/or migration and integration of additional C2 solutions. GCCS-M Increment 2 modernization will be limited to bug fixes, informational assurance–related fixes and patches, and alignment with commercial off-the-shelf/government off-the-shelf (COTS/GOTS) hardware and software technology refreshes.

2. Since its inception, the C2RPC science and technology initiative has been continuously gathering and refining concepts for improved C2. The resultant prototype system defines a new style and approach to C2. This alternative will augment GCCS-M Increment 2 capabilities with productized C2RPC capabilities and newly developed capabilities following the style and approach defined by the C2RPC. This system will provide

a much wider range of C2 capabilities from the MOCs to the tactical level. New software will be installed and integrated with GCCS-M on afloat and ashore GCCS-M infrastructures. In either case, the end user will experience a seamless, integrated C2 system. Modernization will be limited to bug fixes, IA-related fixes, and refreshes.

3. This alternative requires development of a new system to satisfy maritime C2 requirements as defined in the initial MTC2 guidance from the maritime operational level to the tactical edge. It will maintain backward compatibility with existing GCCS-M systems. This alternative builds on the intelligence community data analytic cloud capability, specifically technology demonstrated by National Security Agency called *Ghost Machine* and the Distributed Common Ground System-Army (DCGS-A) program,[1] and ONR Magic Mirror demonstrations.[2] Continuous modernization will include bug fixes, IA-related fixes and patches, and alignment with COTS/GOTS hardware and software technology refreshes.

4. Development of a new system to satisfy MTC requirements as defined in the initial MTC2 guidance from the maritime operational level to the tactical edge, while maintaining backward compatibility with existing GCCS-M systems and transitioning applicable C2RPC capabilities. This alternative adds the concept of the Naval Tactical Cloud (NTC), which places the data analytic cloud afloat, along with providing significant additional storage space for afloat units to be preloaded with historical C2 information. The NTC will allow for continuous synchronization between afloat tactical units and the shore data analytic node(s). New software development will follow agile development techniques, with continuous end-user involvement and responsiveness to fleet requirements. Continuous modernization will include bug fixes, IA-related fixes and patches, and

---

[1] Ghost Machine is the National Security Agency's cloud stack.

[2] Magic Mirror is "a 24–7 command and control capability to monitor and assess the Intelligence Architecture" (ONR, 2011).

alignment with COTS/GOTS hardware and software technology refreshes.

Figure 2.1 depicts the four alternatives and ascribes a short name to each.

Alternatives one and two manage a large number of interfaces to data sources through legacy stovepipes. Alternatives three and four migrate the data structure into a new cloud architecture. Alternatives two, three, and four bring along software modernization schemes to improve the user interface, among other areas.

Figure 2.1
Alternatives for MTC2 Range from Legacy to Cloud

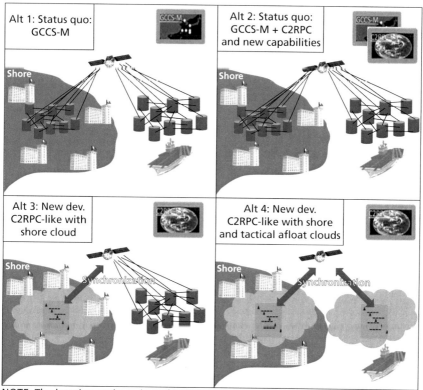

NOTE: The location and number of buildings have no importance in the figure.
RAND RR1383-2.1

For the purposes of modeling and simulation as well as discussion, the key differentiators between the alternatives are summarized in Table 2.1.

A critical assumption for this analysis is the interpretation of the statement "and newly developed capabilities following the style and approach defined by the C2RPC" from the Alternative 2 description.

Table 2.1
Breakdown of Alternatives

	Alternative 1	Alternative 2	Alternative 3	Alternative 4
Description	Status quo GCCS-M Inc. 2	GCCS-M Inc. 2 and production C2RPC	New "C2RPC-like" and shore cloud	New "C2RPC-like," shore cloud, and afloat cloud
Development needed	Limited: some bug fixes, IA patches, and alignment with CANES	Some: significant to productize C2RPC and add new tool support, bug fixes, IA patches, and alignment with CANES	Significant, new tools, architecture, and data structures	Significant, new tools, architecture, and data structures
Tool support for new MTC2 mission tasks	Very limited	Significant	Significant	Significant
Interoperability	Limited data sources	More data sources and intensive to manage	Many data sources and flexible management	Many data sources, flexible management
Data structure	Legacy stovepipes	Legacy stovepipes and some enhancements	Enhanced afloat and NTC RI cloud ashore	NTC RI cloud ashore and afloat
CANES SOA		✓ Limited use of afloat core services	✓ Afloat core services	✓ Afloat core services
Legacy shore HW	✓	✓	✓ (until phased out)	✓ (until phased out)
CANES afloat HW	✓	✓	✓	✓

The resulting assumption is documented in Table 2.1 as "significant" tool support for new MTC2 MTs. The presumption is that a considerable amount of time and money will be spent to achieve new capabilities with Alternative 2. This was the assumption carried forward in the performance, cost, and risk analyses. Out of interest to the program office, a different interpretation of the statement was additionally considered in the cost analysis. The second interpretation indicates that there is not significant new tool support for new MTs. This assumption makes Alternative 2 weaker in performance, but also less costly. The second interpretation is referred to in this report as "2–" (2 minus additional tool support).

CHAPTER THREE

Determination of Effectiveness Measures

3.1. Mission Tasks

MTs were derived from the C2 JIC and initial MTC2 guidance. Four capabilities were selected in the modeling and simulation to keep the effort within scope, given the short turnaround time for the AoA. However, all eight capabilities, and thus MTs, were considered in the tabletop exercise. AoA guidance indicates that each MT should have at least one measure of effectiveness (MoE), and each MoE at least one measure of performance (MoP) (Office of Aerospace Studies, 2008 and 2010). The MoEs also came from the C2 JIC and initial MTC2 guidance, and were selected based on their relevance to the MT. Figure 3.1 details the hierarchy and MTs.

3.2. Measures of Effectiveness

MoEs (Table 3.1) are derived from MTs that are traceable to MTC2-applicable Joint Capability Areas (JCAs), developed by the Joint Staff. In order to determine the threshold requirements for the measures, the study team reviewed the OPLANs for the scenario outlined in the study plan, observed relevant fleet operations and exercise, and interviewed operators integral to carrying out missions in the prescribed scenario. Recommended threshold values used in the grading criteria were drawn from the C2 JIC and initial MTC2 guidance.

Figure 3.1
Linking Mission Tasks to Measures

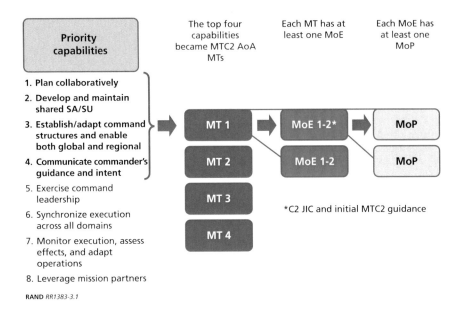

RAND RR1383-3.1

3.3. Measures of Performance

MoPs are derived from MoEs (Table 3.2). This ensures that MoPs are also traceable back to MTs and JCAs.

3.4. Measure Selection Process

Both effectiveness and performance measures were selected based on an assessment of

1. their existence as a metric relative to a MT
2. their ability to be modeled and captured given the suite of modeling capabilities available to the team.

The second criteria requires knowledge of modeling capabilities, alternatives, and technologies involved to derive an understanding of

Table 3.1
List of Measures of Effectiveness

Mission Tasks	Measures of Effectiveness
1. Plan collaboratively	1-1: Percentage of data and information that is interoperable between joint, interagency, intergovernmental, and multinational (JIIM) partners
	1-2: Timeliness of planning information dissemination to JIIM
2. Develop and maintain shared SA and understanding	2-1: Percentage of friendly force locations that are accurate, timeliness as a proxy for accuracy[a]
	2-2: Completeness of COP (percentage of forces)
	2-3: Percentage network use (average across units)
3. Establish/adapt command structures and enable both global and regional collaboration	3-1: Percentage of C-2 data that is interoperable across tactical/operational echelons
	3-2: Likelihood of being able to reconfigure to dynamic mission requirements and returning to steady state
4. Communicate commander's intent and guidance	4-1: Percentage of time orders are received in time to conduct the task/mission
	4-2: Percentage of the commander's plans received by appropriate maritime personnel

[a] The concept is that if you are collecting and processing many friendly locations over time, you will better be able to fuse and improve position accuracy.

the level of difficulty in adjusting the models to capture the metric with enough fidelity in an attempt to discriminate between the alternatives. In some cases, the team was ultimately successful discriminating between the alternatives for MTs one and three, but less successful with MT two. In the latter case, the selected scenario had a significant impact on the lack of compelling results.

An initial set of selected MoEs and MoPs were vetted with PMW 150 staff and technical experts before deciding on the final set used in the analysis. MoP priorities were not established. When aggregating MoP performance to determine MoE performance in the analysis results, each MoP was treated with equal weight.

Table 3.2
List of Measures of Performance

Measures of Effectiveness	Measures of Performance
1-1: Percentage of data and information that is interoperable between JIIM partners	Percentage of DoD C2 data/information that is interoperable
	Percentage of U.S. interagency C2 data/information that is interoperable
	Percentage of five-eye partner C2 data/information that is interoperable
1-2: Timeliness of planning info dissemination to JIIM (define some piece of planning info and follow it through a workflow)	Time for planning information dissemination between joint partners
	Time for planning info dissemination to interagency
	Time for planning information dissemination to multinational partners
2-1: Percentage of friendly force locations that are accurate, timeliness as a proxy for accuracy	Number of red forces on COP (plotted over time)
	Number of blue forces on COP (plotted over time)
	Number of white ships on COP (plotted over time)
2-2: Completeness of COP (percentage of forces)	Neutral force location error (average)
	Blue force location error (average)
2-3: Percentage of available bandwidth consumed	Data rate over links during mission
3-1: Percentage of C2 data that is interoperable across tactical/operational echelons	DoD data interoperability
	US interagency interoperability
	Multinational data interoperability
3-2: Likelihood of being able to reconfigure to dynamic mission requirements and returning to steady state	(Used in the tabletop exercise only, therefore no MoP is specified)
4-1: Percentage of time orders are received in time to conduct the task/mission	Timeliness of promulgation of guidance
	Time to conduct the task/mission
4-2: Percentage of commander's plans received by maritime commanders	Timeliness of promulgation of guidance

Table 3.3
Grading Criteria Used for Aggregate Analysis

MT	MoE	Red	Yellow	Green
1	1-1: Percentage of data and information that is interoperable between JIIM partners	< 50%	50–90%	> 90%
	1-2: Timeliness of planning information dissemination to JIIM[a]	< 50% within one hour	50–90% within one hour	90% within one hour
2	2-1: Percentage of friendly force locations that are accurate, timeliness as a proxy for accuracy[b]	< 50%	50–90%	> 90%
	2-2: Completeness of COP (percentage of forces)			Acceptable
	2-3: Percentage of network use	> 40%	20–40%	< 20%
3	3-1: Percentage of C-2 data that is interoperable across tactical/operational	< 50%	50–90%	> 90%
	3-2: Likelihood of being able to reconfigure to dynamic mission requirements and returning to steady state	< 25% gain over status quo	25–50% gain over status quo	> 50% gain over status quo
4	4-1: Percentage of time orders are received in time to conduct the task/ mission	< 50% within one hour	50–99.9% within one hour	99.9% within one hour
	4-2: Percentage of the commander's plans received by appropriate maritime personnel	< 80% within five minutes	80–90% within five minutes	99% within five minutes

[a] Disseminate with one hour of mission start in the case of PR, four hours of the Commander's Update Brief.
[b] A force location is determined to be accurate in 2-1 if it can be distinguished from other vessels.

Finally, the AoA study guidance indicated that results needed to be summarized in a red, yellow, green format. Table 3.3 shows the grading criteria use to determine the colors for MoE.

The green metrics were developed based on C2 JIC minimum values. Other categories were based on available data in the JIC and initial MTC2 guidance, and augmented with a reasonable discriminating range. MoE 2-2 did not prove discriminating in the analysis, and all alternatives were scored green. The result of the analysis for the top four priority MTs can be found in Chapter Five.

CHAPTER FOUR

Methodology

4.1. Models, Simulations, and Source Data

The performance modeling used three different approaches to analyze the alternatives. Figure 4.1 highlights the three approaches along with a brief description of each.

The MoEs described in Chapter Three were mapped to an analytic approach because not all of the MoEs were easily captured by one or multiple approaches. Other MoEs that addressed interoperability of reconfiguring mission requirements were better studied using

Figure 4.1
Assessment Analytic Approaches

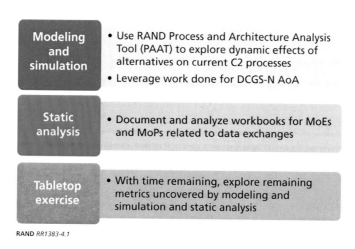

RAND *RR1383-4.1*

Figure 4.2
Mapping of MoEs to Primary Analytic Approach

	MoE	Primary analytic approach
1	1-1: Percentage of data and information that is interoperable between JIIM partners	Static analysis
	1-2: Timeliness of planning information dissemination to JIIM	Modeling and simulation
2	2-1: Percentage of friendly force locations that are accurate, timeliness as a proxy for accuracy	Modeling and simulation + Static analysis
	2-2: Completeness of COP (percentage of forces)	Modeling and simulation
	2-3: Percentage of network use	Modeling and simulation
3	3-1: Percentage of C-2 data that is interoperable across tactical/operational echelons	Static analysis
	3-2: Likelihood of being able to reconfigure to dynamic mission requirements and returning to steady state	Tabletop exercise
4	4-1: Percentage of time orders are received in time to conduct the task/mission	Modeling and simulation
	4-2: Percentage of the commander's plans received by appropriate maritime personnel	Modeling and simulation

RAND *RR1383-4.2*

static, workbook-based analysis or through the tabletop exercise. Figure 4.2 identifies the mapping.

Modeling and Simulation

PAAT was used for the performance-modeling analysis. The model was developed at RAND to support the Distributed Common Ground System-Navy (DCGS-N) Increment 2 Analyses of Alternatives. As shown in Figure 4.3, it integrates sensors, platforms, network links,

Figure 4.3
Overview of PAAT Model

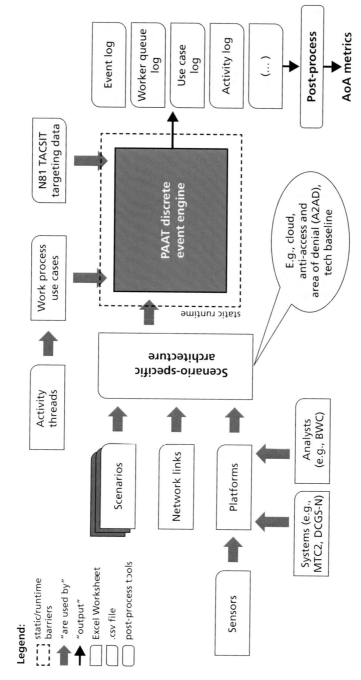

analysts, and systems into dynamic workflows executed within the context of a scenario.

Currently, PAAT exists as a Microsoft Visual Studio project written in Visual Basic .NET. Data tables are stored in a relational structured query language (SQL) database. As depicted in Figure 4.3, there are multiple types of sensors that are then used to compose multiple types of platforms. Additionally, there are network links and scenarios (and vignettes) that, along with the platforms, are used to define a scenario-specific architecture. Many different architectures can be defined, though the baseline in PAAT is from the Navy baselines.

When instantiated, the discrete event engine reads in all of the database inputs to include user-defined inputs about how the scenario should be run. The analysis options may also specify the engine to use maximum data-collection sizes or a random level of background traffic, among others. The engine then simulates the scenario as a collection of events. Each event is logged so that the user can examine what happened during the scenario, if necessary. The key concept in the engine is the data object that is created by the platform's sensor and is then modified and queued as necessary as it traverses the architecture, making the model data-centric.

Two other tools are used to support PAAT. The first is an agent-based tool called Map Aware Non-Uniform Automata (MANA), a combat and C4I, surveillance, and reconnaissance model developed by the New Zealand Defence Technology Agency. MANA is plotted with the position information of blue, red, white, and green forces, depending on the scenario or vignette, and outputs sensor detections to generate a significant portion of the intelligence, surveillance, and reconnaissance data moving through the PAAT architecture. The second tool to support PAAT is a simple, batch-run workbook to enable users to define a series of simulation excursions and "data farm" broad designs of experiments. Parameter distributions for the different alternative options are entered into a design of experiments, which is exercised in a Monte Carlo method. Each excursion within the design of experiments has a different set of initial inputs and is exercised a certain number of times (i.e., runs), where each run uses a series of random seed numbers that are consistent across each excursion.

Workflows

Workflows can be created from many different perspectives and levels of abstraction. Some examples include providing a commander's update brief, developing a personnel recovery mission, providing SA, managing common operating picture tracks, analyzing and processing full-motion video, and exploiting naval organic tactical image. Some examples of the types of insight the model can provide include assessment of network use, timeliness of information dissemination, the state of the common operating picture across the fleet, and identification of workflow bottlenecks.

The workflows or processes were developed based on Fleet Forces Command and U.S. Navy 6th Fleet MOC process diagrams. They were further augmented in discussions with 7th Fleet–watch floor staff. Work processes are initially placed into groups-based functional areas based on the commander's decision cycle and its supporting areas. Figure 4.4 shows the functional areas under consideration.

Each functional area has use cases that describe information inputs, analyst activities, and outputs while interacting with the system (e.g., MTC2). Use cases are described in sequence diagrams, showing a time sequence of events by people, systems, communication paths, and data. The workflows are further broken down into activity threads. Activity threads define the type of information transfer (e.g., face to face,

Figure 4.4
Functional Areas Derived from MOC Operational Views

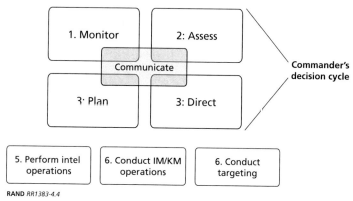

RAND RR1383-4.4

Figure 4.5
Method of Encoding Workflows

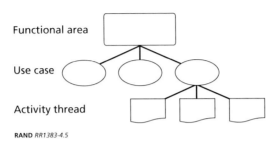

Functional area

Use case

Activity thread

RAND *RR1383-4.5*

email, download, etc.), and map directly to how an analyst is queued with tasks in the model. Figure 4.5 highlights the hierarchical nature of encoding workflows into PAAT.

Table 4.1 shows the functional areas, use cases, and activity threads that exist in the model to date.

Because of the short time constraints of the study, three workflows were selected to represent C2 activities in the Navy: the commander's update briefing, planning and executing a personnel recovery mission, and manage COP and tracks. Figure 4.6 describes the Commander's

Table 4.1
Use Cases and Threads in PAAT

Functional Area	Use Cases	Activity Threads
Monitor	Limited available documentation	
Assess	1	4
Plan	1	7
Direct	6	22
Perform intelligence operations	21	66
Conduct information management/ knowledge management operations	Useful in other performance analyses[a]	
Conduct targeting	3	12
Total	32	111

[a] The architecture of the PAAT model is not amenable to assessing these kinds of operations.

Figure 4.6
Prepare Commander's Update Brief Workflow

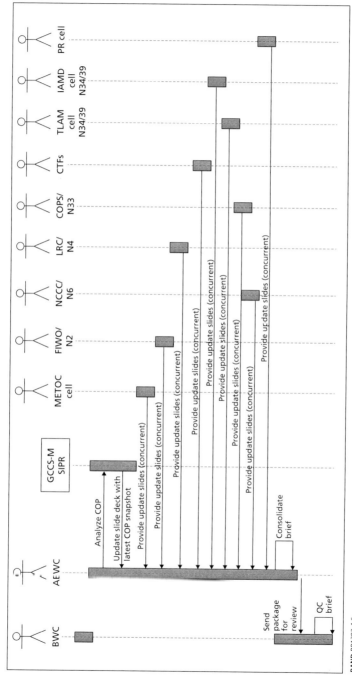

update brief workflow, the importance of which was stressed by the fleet to the AoA team. In many cases, it can be very time consuming for staff to put together a daily brief, in 6th fleet, 4–6 hours on average.

Because of the complex and nested nature of activities in this workflow, the actors perceived to be most important to the mission and their aggregate tasks were considered in the model. As the figure shows, the workflow is highly concurrent and consists of many organizations engaged in cross-cutting and stove-piped information analysis, followed by distillation into a format, nearly always Microsoft Power-Point, to be briefed to the fleet commander. The Assistant Battle Watch Captain is usually responsible for consolidating actor inputs and updating the SA map graphics, which varies day to day and from MOC to MOC, and the Battle Watch Captain reviews the final product and disseminates it accordingly.

In contrast to the commander's update brief, planning for personnel-recovery mission has more-aggressive time demands on staff. It too is highly concurrent and makes use of many different personnel and systems, depending on the mission. In our modeling, we assumed a downed allied aircraft in a variety of different circumstances to scope the workflow. This mission also examines a different aspect of the commander's decision cycle (that of "plan") than the aspect outlined in the update brief (which is primarily "assess"). Figure 4.7 shows the personnel recovery workflow.

Finally, the manage COP tracks workflow was used to exercise functions related to managing the COP. Historically, this is the primary function of C2 support tools, with GCCS-M being the exemplar. Figure 4.8 depicts the workflow.

Using a workflow that models track management enables the model to collect data on the state of the COP given the other intelligence-related tasks that are taking place to process, exploit, and disseminate information before it becomes a track.

Figure 4.9 highlights the mapping of MTs to workflows and the relevant MoEs.

It is important to note that each workflow is considered and potentially altered based on the assumptions for each alternative. For example, since there are four alternatives in this assessment, there are

Figure 4.7
Develop Personnel Recovery Mission Workflow

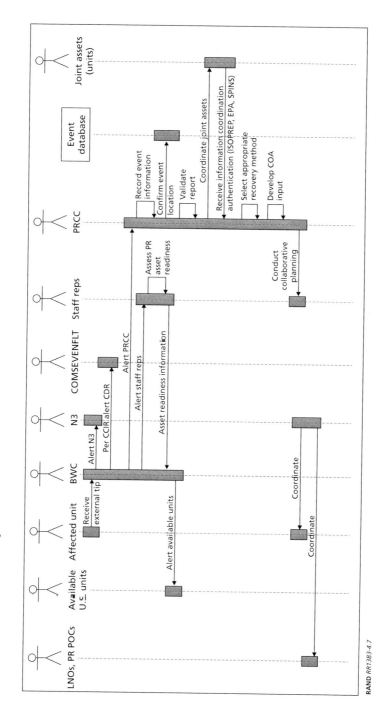

Figure 4.8
Manage the COP and Tracks Workflow

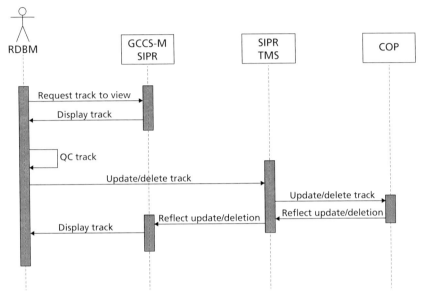

RAND *RR1383-4.8*

Figure 4.9
Mapping of Mission Tasks to Workflows

Mission tasks	Provide CDR's update	Personnel recovery	Manage COP tracks
1. Plan collectively	✔ MoE 1-2	✔ MoE 1-2	
2. Develop/maintain shared SA and understanding	✔ MoE 2-3	✔ MoE 2-3	✔ MoEs 2-1, 2-2, 2-3
3. Establish/adapt command structures			
4. Communicate CDR's intent and guidance		✔ MoEs 4-1, 4-2	

RAND *RR1383-4.9*

four variants of the commander's update brief. This is a main way in which the alternatives are discriminated quantifiably.

Static Analysis

The static analysis was primarily used for MoE 1-1, percentage of data and information that is interoperable between JIIM partners, and 3-1, percentage of C2 data that is interoperable across tactical/operational echelons. The alternatives were examined against 117 data sources derived from GCCS-M, C2RPC, and the scenario. For MoE 1-1, we considered the sources against the relevant JIIM partners from the scenario. For MoE 3-1, we considered the sources against the tactical and operational echelons operating in the scenario to include the CJTF; Component Commands; and task forces, groups, and units. We considered interoperability simply as a binary (1 for yes, 0 for no) about whether the architecture, systems, and software would support a machine-to-machine exchange of information with the data source, with limited manpower required to develop the exchange relationship. The question as to whether data should be shared was not considered, only that it could be shared.

Tabletop Exercise

The tabletop exercise was primarily conducted to assess MoE 3-2, the likelihood of being able to reconfigure to dynamic mission requirements and returning to steady state, as it was not an MoE that was easily captured in the previous two analytic approaches. However, all MoEs across all MTs were considered during the exercise. The tabletop exercise supported the findings of the modeling and simulation where the MoEs overlapped. Extended discussion of the approach and results is in the Appendix.

4.2. Cost-Analysis Approach

This section provides a summary of the major life-cycle cost ground rules and key assumptions used across the development, deployment, and sustainment phase cost elements. They were used as the primary basis for estimating the 50-percent confidence-level estimates in section 5.2 in Chapter Five). Additional assumptions made for the higher 80-percent confidence-level estimates are also summarized in Chapter Five for the mitigation of known risks. See section 5.3 for more details.

Development Costs

The 50-percent confidence-development phase cost estimates for all the MTC2 alternatives is based primarily on software sizing and reuse factors from a source lines of code (SLOC) estimate provided to the team. The estimate covers the PMW 150 software-only development cost. These begin in fiscal year (FY) 2014 with the first incremental release (R1) or FCR-1 and continuing through the fifth release (R5) or FCR-5.

In both appendixes of the document, each incremental release assumes a short development and testing timeline of, on average, approximately 12 months. This applies to all of the MTC2 alternatives through four software incremental releases. This span of time is consistent with PMW 150 assumptions to use and implement a "Rapid Information Technology" acquisition process type of software development activity.

The 50-percent development cost estimate also assumes no software growth in estimated SLOC. Since the MTC2 program of record acquisition was initiated in FY 2012 with funding through FY 2013, the same sunk cost of $7.5 million was added to the development cost estimates for MTC2 alternatives two through four.

In comparison, the development-phase costs for the higher 80-percent confidence level estimates accounts for

- up to an 18 month, on average, software development and testing timeline for each incremental release
- between a 20- to 30-percent increase in potential SLOC growth.

This software growth is based on a separate analysis of the MTC2 Software Support Activity (SSA). We applied code reuse metrics provided to the team by the program office to the updated set of SLOC count data. We used the results of the increased total effective SLOC estimates as the basis for software growth reflected in the higher 80-percent confidence-development cost estimate rather than the 50-percent estimate.

These and other key assumptions and factors driving the 80-percent confidence-level development and the other two life-cycle cost (LCC) phases estimates are summarized below and described in detail in section 5.3 of this report.

In addition to software development costs, the development-phase estimates for the 50- and 80-percent confidence levels also are included the PMW 150 systems engineering estimates for developing MTC2 top-level system specifications and software interface requirements for the CANES hardware stack, virtual memory, data storage, etc. However, all hardware procurement costs for meeting alternatives three and four cloud-based environments are assumed paid for by PMW 120 and are not included in the MTC2 PMW 150 development as well deployment- and sustainment-phase cost estimates.

This also includes potential hardware procurement costs paid for by the PEO C4I and/or the DCGS-N Increment 2 program early on as part of the ONR/PEO C4I NTC Limited Technical Experiments demonstration of Prototype DCGS and MTC2 (Alternative 2 C2RPC and Dynamic C2) Release 1 software deployment on a ship-based platform in the fourth quarter of FY 2014, installed on an ONR-driven hardware stack reference implementation architecture.

Deployment Costs

The deployment costs for alternatives two through four at the 50-percent confidence level all assume that upgrades to CANES hardware will be in place to meet the MTC2 annual fielding objective for installing incremental software (S/W) releases by the FY at each of the planned afloat and ashore sites in the same FY it is made operationally available. This assumes that there will be no added deployment costs for potential delays in planned upgraded CANES platforms expected at these sites.

In comparison, the 80-percent higher confidence-level deployment estimates are based on CANES installation delays by FY provided to the team by the program office.

For both the 50- and 80-percent confidence-level estimates, the deployment-phase costs for Alternative 2 includes S/W platform–site activation, installation, and initial training of first incremental releases FCR-1 across planned ashore and afloat sites. The deployment phase costs for alternatives three and four include S/W platform–site activation, installation, maintenance, operations and support, and initial training from incremental releases (FCR-2 until the final FCR-5 S/W release across planned ashore and afloat sites).

Rather than sending a support team to install and provide update fixes to incremental software releases, the deployment costs estimates are based on being able to remotely upgrade versions of the software to the ashore and afloat sites. For MTC2 alternatives three and four, all the S/W updates or fixes prior to the final FCR-5 incremental release are included as part of the deployment cost estimates. The deployment costs also cover estimates for S/W patches for resolving critical IA issues, bug fixes, etc., prior to the final S/W release.

MTC2 software deployment training costs are based on the assumption that there is no formal schoolhouse training. Deployment training costs through final delivery date (FDD) instead covers:

- preparing instructional materials for Interactive Courseware (ICW) training remotely after each R1 through R5 S/W installation
- student population's time in terms of hourly labor rate going through annual ICW on-site training/certification after installation of Alternative 2 R1 S/W and alternatives three and four R2 through R5 incremental S/W site installations.

We assumed, on average, 32 hours (or between 24 and 40 hours) to complete the ICW training and certification.

For the 50-percent deployment confidence-level estimates, the MTC2 number of billets coincides by FY, with the MTC2 fielding plan objectives across specific sites beginning in

- FY 2015 after Alternative 2 software R1 or FCR1 installation
- from FY 2016 through FY 2020 after alternatives three and four S/W R2 or FCR2 through software R5 or FCR5 installations.

Sustainment (Alternatives One and Two)

GCCS-M Increment 2 sustainment cost is based on program life-cycle cost estimate (PLCCE) provided by PMW 150 from FY 2014 through FY 2030. Alternative 2 contains additional sustainment costs for C2RPC transitioned to GCCS-M Increment 2 beginning in FY 2015 and continuing forward through FY 2030.

Sustainment (Alternatives Three and Four)

Both GCCS-M Increment 2 and MTC2 are sustained simultaneously during MTC2 deployment. Transition from GCCS-M Increment 2 begins on fielding of FCR-2 on group level afloat platform sites and continues through FCR-5 on subs and non-MOC ashore sites.

For the 50-percent sustainment-phase confidence-level estimates, the transition from continuing GCCS-M Increment 2 sustainment cost estimates to phasing in MTC2 alternative three or four is assumed to begin in FY 2018 after S/W installation at group-level afloat sites through FY 2020 after installation within submarines and non-MOC ashore sites. Complete GCCS-M sustainment phase out occurs after MTC2 FDD and when FCR-5 S/W installation on alternatives three or four at the last sites are completed at the earliest in beginning in FY 2021 and continuing out to FY 2030. As stated previously, all prior S/W maintenance costs for fielding earlier FCR-2 through FCR-4 releases are captured as part of the deployment costs for these two alternatives.

Study Limitations/Recommended Next Steps for Quantifying PMW 150 S/W Cost Differences Between MTC2 Alternatives Three and Four

This section describes the study limitations in available data that limited our ability to quantify the differences in the PMW 150 S/W development costs between MTC2 alternatives three and four.

- The S/W development costs we estimated are based on component category-level S/W SLOC estimates and cost-metric values (e.g., reuse factors). We ended up computing S/W development costs based on effective SLOC estimates provided by PMW 150 using the same GCCS-M Increment 2 PLCCE COCOMO II parametric cost model Cost Estimating Relationships as the PMW 150 cost group.
- The S/W components were not mapped functionally to initial MTC2 guidance or the C2 JIC set of capabilities and tasks required. From a completeness perspective, we had no direct way of assessing how well each of the new alternatives two, three, and four met the expected set of MTC2 requirements.
- Furthermore, beyond Alternative 2 FCR-1 productized C2RPC components, there are no common and/or unique S/W tools and application releases specifically identified for Alternative 3 shore-only cloud sites over both Alternative 4 shore and afloat ship-based cloud products.

Going forward, we recommend that, as part of the process of updating the LCC estimates for alternatives three and four, that PMW 150 consider:

- continuing the S/W sizing and metrics assessment effort initiated by the MTC2 SSA organization as an updated, current, and much-needed improved basis for quantifying cost, the level of effort versus the capability alternatives trade-space differences
- breaking out MTC2 SSA's SLOC count estimates by releases (R1 through R5) across MOCs and level afloat ships and submarine sites at S/W comparable component category summary or lower levels as earlier estimates
- determining which S/W applications, tools, and widgets are common or unique to specific MTC2 Alternatives 2, 3, and 4 by first reviewing, using, and, if needed, expanding the list from the January 2013 C2RPC TRA report
- assessing the level of reuse or modification for each software component as an estimated percentage of total SLOC along with esti-

mated new SLOC needed to meet expected functionality mapped to specific JIC and initial MTC2 guidance requirements

• completing the MTC2 cost-assessment template, illustrated in Figure 4.10, for the other software metric values needed

• leveraging off of Space and Naval Warfare Systems Command (SPAWAR) 1.6 cost team's recent independent assessment of the expected SLOC code growth.

4.3. Risk-Analysis Approach

The risk assessment judged each alternative on a number of risk areas and scored them as low, medium, or high risk according to U.S. Air Force AoA handbook standards (Office of Aerospace Studies, 2010). The risk assessment is broken down into multiple subcategories according to the U.S. Air Force AoA handbook guidance, which notes that "the qualitative evaluation of risk consequence will be determined by the criteria shown" (Office of Aerospace Studies, 2010, p. 40). The subcategories were assessed qualitatively by RAND analysts using the criteria, and scores for overall technical, schedule, programmatic, and cost risk were developed.

Background

For the purposes of this report, risk is defined by the probability that something adverse will occur, in addition to the concomitant consequences. This means that a high-risk score for a risk item, such as the NTC deployment schedule, indicates a high likelihood and/or severity of failure on that item's part.

Our method for assessing risk for MTC 2 AoA follows guidelines in the USAF AoA handbook. Risk is calculated on a matrix of probability and consequence and falls into regions of low, medium, and high risk. Figure 4.11 shows the USAF AoA handbook's risk matrix.

The matrix is not quite symmetric. Items with very low probability of occurrence but high consequence if they occur are deemed moderate risk, while items with a very high probability of occurrence but very low consequence are deemed low risk. Outside these two blocks, the risk matrix is symmetric.

Figure 4.10
MTC2 Cost Estimating Assessment Template

Software Component:									
Reviewer					Date:				
Instructions: Perform a Cost Estimate on the Widget(s) below, providing SLOC estimates in the variation designated. Provide data to Cost Estimators so that they may run COCOMO model.									
Use the following metric for defining Assessment: True – Cost calculated, enter amount for widget specified False – Cost unable to be calculated (enter 0)									
Component list	Total SLOC	New	Modified	Rehosted	Translated	Verbatim	COTS Integrated	Auto Generated	Cost (or 0 if cannot calculate)
					Overall Assessment Criteria: (based on % of widgets successfully estimated) 80%–100% - Green 60%–80% - Yellow <60% - Red		Total:		
Assessment: (R/G/Y based on assessment criteria) - include comments which reflect ambiguity on code analyzed.									

SOURCE: McDonnell, 2014.
NOTES: Definitions: new = code written from scratch. No reuse of design or legacy code; modified = existing code that was altered, enhanced, adapted, etc., to be used in current project; rehosted = code that was originally developed for a different hardware platform or operating system that was converted to be used in the current hardware platform or operating system environment; translated = code originally developed in one language that was converted into a new language to be used in the current project; verbatim = preexisting code that was reused in its entirety without any modifications; COTS integrated = code that was written to interface a COTS software package with other custom and/or COTS software, sometimes referred to as glue code; auto-generated = code that was produced automatically by an auto-generation program (i.e., Microsoft FrontPage–generating HTML code).
RAND RR1383-4.10

Figure 4.11
USAF AoA Handbook Risk Matrix

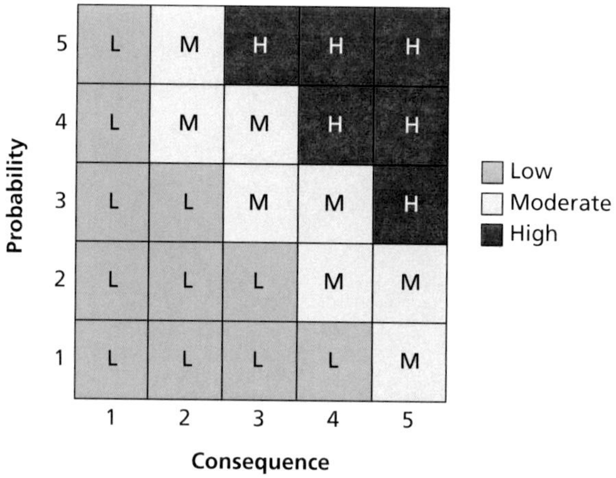

The USAF AoA Handbook offers further clarity on what the various levels of probability and consequence represent. While risk probability is identical for all risk types, risk consequence has slightly different meanings for the three different areas of risk (technical, schedule, and cost). Figures 4.12 and 4.13 explain risk probability and consequence.

A high-risk probability score indicates that the system component being examined will likely fail in its intended use or intended schedule or that the negative external action being examined is likely to happen.

Risk consequence indicates the severity of the impact on the MTC2 system if the risk item under discussion fails to perform as desired or be implemented on time or if the external negative action under discussion succeeds.

Per the study plan, we grouped risk into three categories: performance/technical,[1] schedule, and programmatic. In the course of performing the cost analysis, we also compiled cost risks for each alternative. Those risks also are included in this chapter.

---

[1] Hereafter referred to as *technical risk*.

Figure 4.12
Risk Probability

Level	Likelihood	Probability (*P*) of occurrence (%)
1	Not likely	$P \leq 20$
2	Low likelihood	$20 < P \leq 40$
3	Likely	$40 < P \leq 60$
4	Highly likely	$60 < P \leq 80$
5	Near certainty	$P > 80$

Increasing risk (arrow pointing down, left of table)

SOURCE: Office of Aerospace Studies, 2010, Table 8-1.
RAND *RR1383-4.12*

Technical risk is based on critical technology elements associated with each alternative. Per the study plan's guidance, technical risk assesses

1. integration of legacy systems, COTS, and GOTS
2. development and integration of new applications
3. compliance with an enterprise framework constrained by the potential infrastructure providers
4. integration with cloud architectures.

The study plan defines schedule risk as the likelihood of completing development, integration, and operational testing activities in time to deliver capabilities from 2014 to 2018 and achieve FDD within five years of program initiation.

Programmatic risks are not specifically called out in the study plan, but during the course of the analysis, we determined that this category, which was considered in the original project description, should be considered as a separate risk category. Programmatic risks arise from programmatic interdependencies that are critical to each alternative's ability to perform stated MTC2 tasks. Interdependencies can include strategic, operational, and tactical C2 systems, Global Information Grid services, shared access to Navy data sources, and coalition systems.

Figure 4.13
Risk Consequence

Level	Performance/technical	Schedule	Cost
1	Minimum consequences to technical performance but no overall impact to the program success.	Negligible schedule slip.	**Pre-MS B**: ≤5% increase from previous cost estimate. **Post-MS B**: Limited to ≤1% increase in Program Acquisition Unit Cost (PAUC) or Average Procurement Unit Cost (APUC).
2	Minor reduction in technical performance or supportability can be tolerated with little or no impact on program success.	Schedule slip, but able to meet key dates (e.g., PDR, CDR, FRP, FOC) and has no significant impact to slack on critical path.	**Pre-MS B**: >5% to 10% increase from previous cost estimate. **Post-MS B**: ≤1% increase in PAUC/APUC with potential for further cost increase.
3	Moderate shortfall in technical performance or supportability with limited impact on program success.	Schedule slip that impacts ability to meet key dates (e.g., PDR, CDR, FRP, FOC) and/or significantly decreases slack on critical path.	**Pre-MS B**: >10% to 15% increase from previous cost estimate. **Post-MS B**: >1% but <5% increase in PAUC/APUC.
4	Significant degradation in technical performance on major shortfall in supportability with moderate impact on program success.	Will require change to program or project critical path.	**Pre-MS B**: >15% to 20% increase from previous cost estimate. **Post-MS B**: ≥5% but <10% increase in PAUC/APUC.
5	Severe degradation in technical /supportability threshold performance will jeopardize program success.	Cannot meet key program or project milestones.	**Pre-MS B**: >20% increase from previous cost estimate. **Post-MS B**: ≥10% increase in PAUC/APUC danger zone for significant cost growth and Nunn-McCurdy breach.

Increasing risk

SOURCE: Office of Aerospace Studies, 2010, Table 8-2.
RAND *RR1383-4.13*

Cost risk is the uncertainty associated with costs related to each alternative. It is not, however, a measure of the overall cost of an alternative. The absolute cost of each alternative is not considered here, only the uncertainty in what that absolute cost might be.

Chapter Five describes how these various risk categories are used to find an overall risk score for each alternative.

Execution

We began risk assessment by following the study plan guidance and USAF AoA Handbook definition of risk. These two sources were used as the definitive guides on risk definition and classification.

We enumerated many technical, schedule, programmatic, and cost risks. Initially, a large number of risk elements were derived for MTC2. Through multiple iterations, we combined similar risks and refined risk-element descriptions to a final list of 15 risk elements. The risks items were distributed into the appropriate categories and subcategories using study plan and USAF AoA Handbook guidelines.

Each risk element was scored by an analyst against each alternative (when appropriate) on risk probability and risk consequence. Analyst scores were then distributed to other RAND analysts for revisions. This process continued until consensus scores were reached.

To score each risk category, analysts examined the applicable risk elements for each alternative. Chapter Five examines scoring for each risk category using the methodology described in this section.

Analysis Results

5.1. Performance-Modeling Results

First, we will explore the performance of two of the workflows explored in the analysis: Commander's update brief preparation and developing a personnel recovery mission. There are at least two dimensions in which performance can be considered: the quality of the product delivered and the timeliness of delivery. Our analysis considers timeliness, as quality is particularly more difficult to assess not only in general, but with modeling and simulation.[1] Figure 5.1 shows the performance of the alternatives in preparing the Commander's update brief.

It is natural to consider duration as a metric for this workflow; however, in reality, the update brief is always generated on time because the brief must be delivered at its regularly scheduled time. For the purposes of the analysis, there was no forcing function to get it done on time, so it can be assumed that, in the model, the analysts were forced to start their analysis earlier in the day in order to finish it on time.

In general, Alternative 4 performs well because of availability of local information. This is due to the fact that, in Alternative 4, the tactical afloat cloud stages more information for staff afloat, minimizing the amount of time they need to search and pull information from ashore nodes (i e , Alternative 3). PAAT can achieve this level of analysis because it models the network dynamically, so delays in download

---

[1] To our knowledge, the Navy gives no guidance on the usability, readability, or completeness of content within a Commander update briefing. This is likely at the commander's discretion, and its study, though important, was outside the scope of this research.

Figure 5.1
Performance of Preparing the Commander's Update Brief Workflow

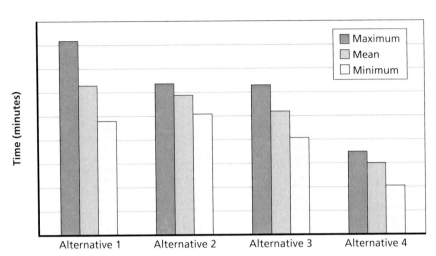

NOTE: Axis values have been abstracted.
RAND *RR1383-5.1*

manifest as delays in the workflow. The model showed that based on the amount and types of requests for information to off-board nodes, particularly ashore, there are areas in the workflow that are sensitive to download and assimilation delays.

Alternative 4 performs well, developing a personnel recovery mission for similar reasons. In at least one excursion, Alternative 3 did worse than even Alternatives 1 and 2. This is due to the stochastic nature of the model, and it was observed that Alternative 3 did better than one and two on average. See Figure 5.2.

The static analysis showed that the cloud options, Alternatives 3 and 4, bring increased potential for interoperability with JIIM partners. Interagency and intergovernmental were considered together. In general, the proposed design of the cloud architecture is more amenable to information exchange than the existing architecture or the options for Alternative 2. See Figure 5.3.

Similarly, Alternatives 3 and 4 also bring high levels of potential interoperability across Navy echelons. Two operational levels were considered, joint task force and Component Commands, as well as three

Figure 5.2
Performance of the Developing Personnel Recovery Mission Workflow

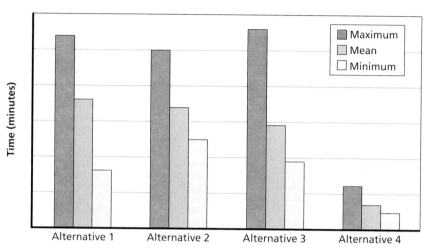

NOTE: Axis values have been abstracted.

Figure 5.3
Percentage of Interoperable Data Sources from MTC2 to Mission Partners

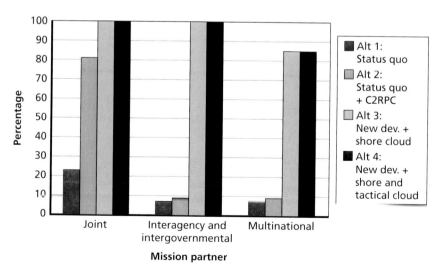

tactical levels, task force, task group, and task unit level platforms. Figure 5.4 shows the results of the analysis.

Overall performance results are shown in Figure 5.5.

In the modeling and simulation of the personnel recovery scenario, planning information was disseminated to JIIM partners four times faster in Alternative 4 than Alternative 3. Also, orders were received in time to conduct the mission significantly more often in Alternative 4 than Alternative 3 and overwhelmingly more often than Alternatives 2 and 1.

The Commander's update brief preparation and planning information was disseminated to JIIM partners one and a half times faster in Alternative 4 than Alternative 3.

Regarding accuracy, completeness, and network use, the differences in the performance between the alternatives for management of the COP were not significant. The ability to meet the requirements for accuracy and completeness for the phase of the scenario simulated was generally good.

Figure 5.4
Percentage of Interoperable Data Sources from MTC2 to Command Echelons

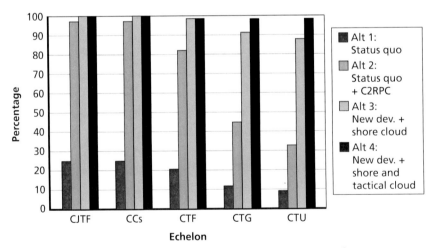

NOTE: CJTF = Combined Joint Task Force; CCs = Combatant Commands; CTF = Command Task Force; CTG = Command Task Group; CTU = Command Task Unit

RAND RR1383-5.4

Figure 5.5
Summary Performance Results for Critical Measures

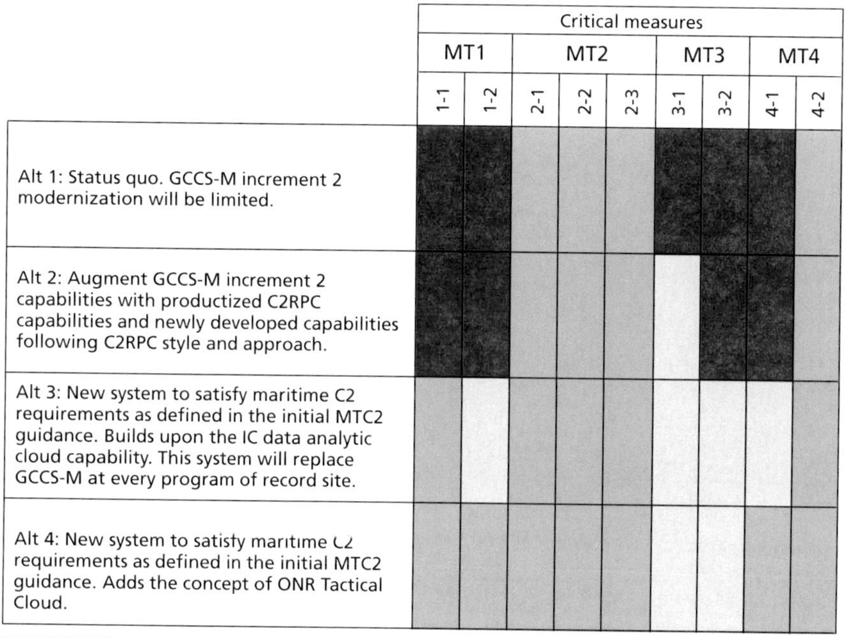

	Critical measures								
	MT1		MT2			MT3		MT4	
	1-1	1-2	2-1	2-2	2-3	3-1	3-2	4-1	4-2
Alt 1: Status quo. GCCS-M increment 2 modernization will be limited.									
Alt 2: Augment GCCS-M increment 2 capabilities with productized C2RPC capabilities and newly developed capabilities following C2RPC style and approach.									
Alt 3: New system to satisfy maritime C2 requirements as defined in the initial MTC2 guidance. Builds upon the IC data analytic cloud capability. This system will replace GCCS-M at every program of record site.									
Alt 4: New system to satisfy maritime C2 requirements as defined in the initial MTC2 guidance. Adds the concept of ONR Tactical Cloud.									

RAND *RR1383-5.5*

5.2. Cost-Analysis Results

Table 5.1 describes the 50-percent life-cycle costs for the MTC2 alternatives.

The summary life-cycle cost for each of the four MTC2 alternatives represents the 50-percent (or most likely) confidence level estimates for PMW 150 to either:

• keep the status quo of sustaining GCCS-M Increment 2 software through FY 2030 listed as Alternative 1
• add productized C2RPC and deploying incremental releases at GCCS-M ashore and afloat sites from MOCs to tactical-level afloat sites and then sustaining the additional software within GCCS-M Increment 2 as Alternative 2-

Table 5.1
Life-Cycle Cost Results (in FY 2013 $M)

Alternatives	Development Cost (FY 2014– 2020)	Deployment Cost (FY 2014– 2020)	GCCS-M Sustainment (Transition) Cost (FY 2014– 2020)	MTC2 Sustainment Cost (FY 2021– 2030) (3)	Total LCC (FY 2014– 2030)
1: Status quo GCCS-M Inc. 2	N/A	N/A	$212 Avg. Annual Cost: $30.3	$195 Avg. Annual Cost: $19.5	$407
2: GCCS-M Inc. 2 + Productized C2RPC + New Compatibilities	$36	$12	$212 Avg. Annual Cost: $30.3	$204 Avg. Annual Cost: $20.4	$464
3: New "C2RPC- like" + Shore Cloud	$107	$22	$191 Avg. Annual Cost: $27.3	$39 Avg. Annual Cost: $3.9	$359
4. New "C2RPC- like" + Shore Cloud + Afloat Cloud	$107	$22	$191 Avg. Annual Cost: $27.3	$39 Avg. Annual Cost: $3.9	$359

NOTE: Estimates are at 50-percent confidence level in constant FY 2013 dollars. We used the same escalation indices consistent with PMW 150's GCCS-M Increment 2 PLCCE. We also used the same work breakdown structure from the GCCS-M Increment 2 PLCCE "GCCS-M_PLCCE_17September2012_iCRB-with 1K risk run," 2012.

- add productized C2RPC plus the cost of adding a modest amount of "newly developed capabilities following the style and approach defined by C2RPC"[2] to GCCS-M afloat- and shore-based sites and then sustaining the additional software within GCCS-M Increment 2 as Alternative 2–
- add productized C2RPC and other software with the cost for developing and deploying this incremental release with a modest amount of new capabilities to afloat- and shore-based sites,

---

[2] As defined for our report, guidance for all the alternatives, including our report's specific context for defining Alternative 2B, is taken from the verbiage describing the new capabilities cited in the U.S. Navy SPAWAR C4I PEO (MDA)-approved "Maritime Tactical Command and Control (MTC2): Analysis of Alternatives (AoA) Study Plan," facsimile provided to authors by the U.S. Navy, January 15, 2013, approved on February 7, 2013.

then sustaining the additional software within GCCS-M Increment 2 as Alternative 2
- develop, deploy, and transition from sustaining GCCS-M to a new "C2RPC-like" shore sites cloud-based system (Alternative 3) or a combined "C2RPC-like, cloud-based" system across shore and afloat site (Alternative 4).

Alternatives 2 through 4 deployment cost includes S/W platform/site activation/installation, S/W maintenance, fleet support tiger teams, and Interactive Courseware Training of incremental releases (R1 through R5) across planned ashore and afloat ship-based sites through FY 2020.

For Alternatives 3 and 4, GCCS-M Increment 2 sustainment cost based on beginning MTC2 transition in FY 2018 with group afloat ship-based sites through FY 2020 with submarines and non-MOC ashore sites. Alternatives 3 and 4 sustainment cost based on complete phase out of GCCS-M at all sites, with FDD beginning in FY 2021.

These estimates do not include contingency costs associated with input uncertainties such as S/W sizing or SLOC growth and the impacts of mitigating known technical- and programmatic (schedule)-related risks.

5.3. Risk Results

The analysis showed Alternative 4 to be the least technically risky option. It is more capable than Alternative 3 and much more capable than Alternatives 1 and 2 at fulfilling technical requirements for MTC2. Alternatives 1 and 2 are least risky when examining schedule and programmatic risks. They have lower schedule risks because Alternatives 3 and 4 are introducing a number of new technologies such as NTC and Agile Core Services (ACS)[3] and face risks from these programs not arriving on time. Alternatives 3 and 4 have higher

---

[3] ACS is a component of CANES to provide shared software services to many application programs in an effort to reduce duplication and cost.

programmatic risk due to the possibility of the large volume of data in the cloud and access to that data not being managed optimally. Alternatives 1 and 2 also have low cost risk, though Alternatives 3 and 4 only have moderate cost risk by comparison.

Technical Risk

Technical risk (Table 5.2) is risk relating to the critical-technology elements associated with each alternative. While these risks can be further broken down into technical risk subcategories—such as design, threat, and requirements—the small number of technical risks for MTC2 does not necessitate this.

Alternative 4 shows the lowest technical risk due in large part to its discarding of current architectures in favor of both afloat and ashore clouds as well as their ability to manage a multitude of data sources and users. Alternatives 2 and 3 show slightly more technical risk with issues arising from using legacy stove piped (meaning data only travels along one or very few paths) architecture aboard afloat units. These issues include collaboration tools, disconnected/interrupted/low-bandwidth (DIL) functionality, and ensuring proper data access. Alternative 3 does mitigate some technical risk items with its ashore cloud. The cloud architecture, at least on the ashore side, alleviates issues related to data access and collaboration. But integration of the two different afloat/ashore architectures presents its own challenges. Alternative 1 shows the highest technical risk. The current data architecture is simply not designed or able to handle the increasing number of data sources that will be needed by MTC2 or the robust security solution needed to ensure proper data access.

Overall, Alternatives 1 and 2 show higher technical risk than Alternatives 3 and 4. This is due in large part to their inability to deal with the overwhelming amount of data that MTC2 will need to manage. Alternatives 1 and 2 both use the current stove-piped data architecture, which is less efficient than the other alternatives as a way to manage the increasingly large amount of data sources the Navy relies on. Alternatives 1 and 2 also lack interoperable collaboration and collaborative planning tools for use within the Navy and with JIIM partners. These

Table 5.2
Technical Risks

	Alt. 1		Alt. 2		Alt. 3		Alt. 4	
	Prob.	Cons.	Prob.	Cons.	Prob.	Cons.	Prob.	Cons.
Fail to provide robust afloat non-cloud data integration layer	N/A	N/A	N/A	N/A	3	5	N/A	N/A
Fail to provide significant functionality in DIL environments	4	5	5	4	4	4	1	4
Fail to identify or stabilize the number of additional data sources feeds needed to enable the COP and collaborative planning, sharing, assessment, monitoring tools	4	4	4	3	4	1	4	1
Fail to access, ingest, validate, and index most (Priority 1 and 2) data needed to enable MTC2 requirements	N/A	N/A	N/A	N/A	1	4	1	4
Fail to provide collaboration and collaborative planning tools that interoperate with JIIM partners	3	5	3	3	2	3	2	3
Fail to provide reliable, redundant data feeds to enable high-priority effects chains (including BMD, IAMD, ASW and SUW)	2	5	5	2	4	2	2	2
Fail to provide user-defined data discovery, query, access, and standardized information-management capabilities across multiple security enclaves (DQS)	N/A	N/A	N/A	N/A	3	3	3	3
Fail to provide robust multilevel security solution, access, and sharing of the full range of MTC2-accessible data for Navy and JIIM partners	4	5	4	4	3	4	2	4

Table 5.2—Continued

	Alt. 1		Alt. 2		Alt. 3		Alt. 4	
	Prob.	Cons.	Prob.	Cons.	Prob.	Cons.	Prob.	Cons.
Fail to mature concepts for automating key MTC2 functions (including track management and correlation, presentation management, operational assessment, COA development, mission analysis, mission synchronization, and predictive modeling)	2	5	5	2	4	2	4	2

NOTE: The colors correspond to those found in the USAF AoA Handbook, and are described in section 4.3 of this report.

alternatives do not have automated tools capable of automating any of these tasks and do not allow for new development of such tools.

Alternatives 3 and 4 fare better in this regard because of their reliance on a cloud-based system. Alternative 3's ashore cloud is well prepared to handle many different data sources. The main technical issues arise from the need for a data-integration layer between the shore cloud and the afloat non-cloud systems. Similar to Alternatives 1 and 2, Alternative 3 will likely not function well in DIL environments. A breakdown of low, medium, and high risks for each of the four alternatives is illustrated in Figure 5.6.[4]

Alternative 4 is the most technically capable of the four alternatives. The ashore and afloat cloud design will mitigate the data integration layer and DIL risks that Alternative 3 has because of its lacking of the afloat cloud. Alternative 4's most-pressing technical risk deals with automated functionality, a risk also shared with Alternatives 1–3.

Schedule Risk

Schedule risks are those that deal with problems that could delay the rollout of MTC2. Schedule risks are more severe for Alternatives 3 and 4 because these alternatives rely on a number of cloud-enabling or

_____

[4] The low, medium, and high breakdown is based on the U.S. Air Force handbook risk matrix discussed in section 4.3 of this report.

Figure 5.6
Alternative 1 Has the Highest Technical Risk, While Alternative 4 Has the
Lowest Technical Risk

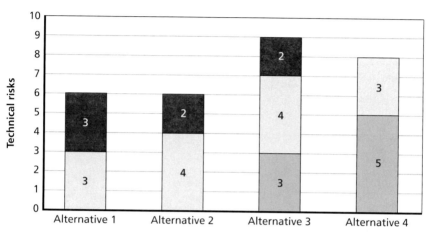

otherwise new systems and tools. Alternatives 1 and 2 lack significant new development and, as such, have little to no schedule risk. Table 5.3 enumerates MTC2 schedule risks.

Alternative 3 relies on on-time fielding of a data-integration layer for cloud-to-non-cloud communication, IDAM, ABAC, and ACS 2.0 for ashore cloud function and ashore-afloat cloud communication.

Alternative 4 relies on the same programs as Alternative 3 as well as NTC. NTC is needed for afloat-cloud function and afloat-ashore communication. Because of low or lack of reliance on these programs, Alternatives 1 and 2 have very few schedule risks, all of which are considered low risk. Figure 5.7 breaks down low-, medium-, and high-schedule risks for MTC2.

Of these cloud-enabling technologies, the data-integration layer is the most risky for Alternative 3. By removing GCCS-M and thereby minimizing stove pipes, the design complexity has shifted from GCCS-M managing many disparate data sources to a shared data-integration layer that the program has less control over. For Alternative 4, the data-integration layer and NTC are by far the most

Table 5.3
Schedule Risks

	Alt. 1		Alt. 2		Alt. 3		Alt. 4	
	Prob.	Cons.	Prob.	Cons.	Prob.	Cons.	Prob.	Cons.
Delay fielding of robust data integration layer and responsibility for creating robust data integration layer	N/A	N/A	N/A	N/A	4	4	4	5
Delay fielding ACS 2.0 (DCGS-N Inc2)	N/A	N/A	2	1	2	4	2	4
Delay fielding enterprise Identity and access management (IDAM)[a] and attribute-based access control (ABAC)[b]	2	1	2	3	2	4	2	4
Delay fielding NTC	N/A	N/A	N/A	N/A	N/A	N/A	4	5

[a] IDAM "is the security discipline that enables the right individuals to access the right resources at the right times for the right reasons." (Gartner, undated.)

[b] ABAC is a logical access control model that is distinguishable because it controls access to objects by evaluating rules against the attributes of the entities (subject and object) actions and the environment relevant to a request." (National Institute of Standards and Technology, 2015.)

critical, both resulting in an inability to field MTC2 with anticipated functionality should they not be ready in time (and both having a high likelihood of not meeting this deadline).

Programmatic Risk

Programmatic risks come from programmatic interdependencies that are critical to each alternative's ability to perform stated MTC2 tasks. Interdependencies can include strategic, operational, and tactical C2 systems, GIG services, shared access to Navy data sources, and coalition systems.

Programmatic risks were not specifically advocated in the study plan, although they were suggested in the original RAND project description. While performing the risk analysis, RAND analysts determined that a number of risks that could be best classified as programmatic were impactful to the four alternatives. Thus, programmatic

Figure 5.7
Alternative 4 Has the Highest Schedule Risk, Followed by Alternative 3,
While Alternatives 1 and 2 Have Low Schedule Risk

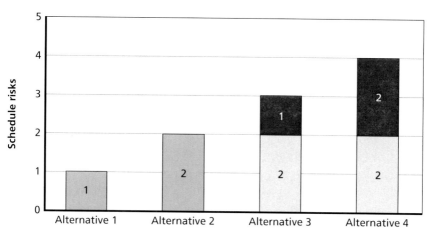

RAND RR1383-5.7

risks were re-added to the analysis. Table 5.4 gives values for MTC2 programmatic risks.

Programmatic risks are most impactful to Alternatives 3 and 4. They have a need for a PEO C4I/Navy/DoD/interagency, cross-PMW cloud architecture governance strategy that is not prevalent in Alternatives 1 and 2.

Table 5.4
Programmatic Risks

	Alt. 1		Alt. 2		Alt. 3		Alt. 4	
	Prob.	Cons.	Prob.	Cons.	Prob.	Cons.	Prob.	Cons.
Fail to institute/enforce PEO C4I/Navy/DoD/interagency, cross-PMW cloud architecture governance strategy	N/A	N/A	N/A	N/A	3	4	3	4
Fail to institute/enforce an effective data strategy (including data ingest and ownership policies)	3	3	3	4	3	5	3	5

While Alternatives 1 and 2 could suffer from not having an effective data strategy in place, Alternatives 3 and 4 would face critical system failure should this occur. Figure 5.8 shows the breakdown of low, medium, and high programmatic risks for the four alternatives.

Like schedule risk, programmatic risk is higher for the cloud-enabled alternatives. This is expected, as introducing new ways of managing data will inevitably lead to risks associated with how that management is performed.

The next section describes detailed risk comparisons. Specifically, it describes the quantitative details of the cost risk differences between the alternatives. Among the option compared are the following:

- low risk Alternative 1 assessment
- the low to moderate risks of Alternative 2A and 2B.

These are compared with Alternatives 3 and 4.

Figure 5.8
Alternatives 3 and 4 Suffer from Greater Programmatic Risk than Alternatives 1 and 2

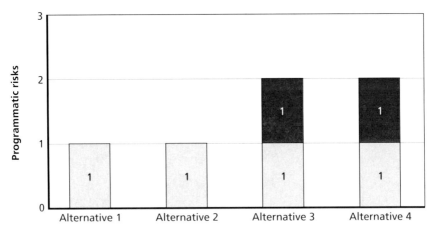

Cost Risk

In addition to the quantitative assessments, dependence on potential schedule risks of external programs delivering updates in the timelines expected could result in impacting the potential cost growth and schedule slips of incremental software releases and installations differently for Alternative 4 compared with Alternative 3. For example, PMW 150's ability to manage and control both the software development costs and development times of incremental releases of Alternative 4 may be more dependent on other things. Specifically, it may be more dependent on the progress, use, and testing of MTC2 software using the latest expected versions of the NTC relative to the implementation of Alternative 3.

The potential schedule risk in the delay in delivering a specific version of NTC over when it was expected may have a greater impact (and higher likelihood). This greater impact is in terms of both:

- increasing the development cost[5]
- increasing the deployment cost.

The increase in deployment cost is due to a potential slip in installation and fielding the designated sites of an Alternative 4 S/W incremental release. This is relative to the potential NTC schedule risk occurring on Alternative 3. See Figure 5.9.

Furthermore, there could be a more-pressing need (more requirement) to complete the external development and installation of a data-integration layer.[6] Specifically for ashore sites, the completion of a data-integration layer for meeting operational requirements is more highly likely to be on the critical path for Alternative 4. Any delays in completing this development effort could have a greater ripple-effect impact on delaying the completion of Alternative 4 incremental S/W releases and installations. In turn, this will result in higher develop-

---

[5] Because of delaying, the testing needed.

[6] At all the designated ashore sites for Alternative 4 prior to completing the installation and fielding of the same MTC2 S/W incremental release as Alternative 3.

Figure 5.9
Overview of Cost Risk

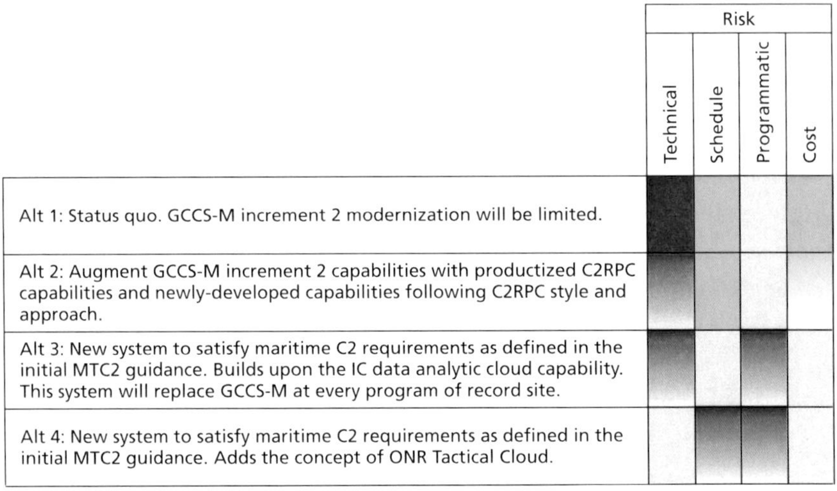

	Risk			
	Technical	Schedule	Programmatic	Cost
Alt 1: Status quo. GCCS-M increment 2 modernization will be limited.				
Alt 2: Augment GCCS-M increment 2 capabilities with productized C2RPC capabilities and newly-developed capabilities following C2RPC style and approach.				
Alt 3: New system to satisfy maritime C2 requirements as defined in the initial MTC2 guidance. Builds upon the IC data analytic cloud capability. This system will replace GCCS-M at every program of record site.				
Alt 4: New system to satisfy maritime C2 requirements as defined in the initial MTC2 guidance. Adds the concept of ONR Tactical Cloud.				

NOTES: Alternative 3 and 4 risks should be mitigated as MTC2 matures. Alternative 1 and 2 risks are not easily mitigated.

RAND *RR1383-5.9*

ment cost growth compared with comparable incremental releases of software for Alternative 3.

Table 5.5 highlights some of the causes for risk and uncertainty in the cost estimates. In particular, the 80-percent level includes estimates for SLOC growth over time.

Although the estimated percent increases in effective SLOC represented a broad range from 46 percent of to 14 times the earlier mission capability-based component estimates, this is only a subset of the eventual number of S/W components and total estimated SLOC that comprises the S/W development effort for productizing C2RPC. In addition, since this representative growth does not represent the unique S/W components for developing Alternative 2 new capabilities, we elected to cap the SLOC growth at an aggregate, yet relatively conservative, increase of 20 percent.

To refine the data inputs for improving the credibility of the cost estimates, we engaged SMEs to identify several cost-discriminating qualitative factors worth evaluating. The SMEs specifically focused on

Table 5.5
Causes of MTC2 AoA PMW 150 LCC Risk/Uncertainty Estimated Variances (FY 2013 $M)

Alternatives	Total LCC Estimates (FY 2014–2030)		Dependent (Closely Coupled) Contributing Factors			
	5C% CL	80% CL	S/W Sizing SLOC Estimates (1)	S/W Incremental Development Span Times	Projected S/W Productivity	Gaps in with Upgraded CANES Hardware
1: Status quo GCCS-M Inc. 2	$407	$407		By definition, sustainment only (no quantifiable cost variance)		
2: GCCS-M Inc. 2 + Productized C2RPC + New Compatibilities	$464	$516 (+11%)	Added 20% SLOC growth based on reusing (as is or modifying) same 8 tools at higher estimates than those used for R1 component-level sizing (2)	Minimal cost impact	Ranged from GCCS-M Inc-2 analog Avg. 367 to 93 Equiv. SLOC/MM	Minimal cost impact

Table 5.5—Continued

Alternatives	Total LCC Estimates (FY 2014–2030)	Dependent (Closely Coupled) Contributing Factors		
3: New "C2RPC-like" + Shore Cloud	$359 $585 (+63%)	Added 30% SLOC growth based on reusing same 5 GCCS-M apps, 3 tools/widgets at higher SLOC as basis for R2 through R5 component-level sizing (2)	Potential increase from 12 to 18 months results in R2 through R5 stretch-out and likely increase in costs through FY 2022 and beyond	Ranged from DCGS-N Inc-2 Cloud CSCI analog Avg. 330 to 178 Equiv. SLOC/MM	Potential stretch-out in deployment cost effort of between six and 24 months over R2 through R5 fielding (3)
4: New "C2RPC-like" + Shore Cloud + Afloat Cloud	$359 $585 (+63%) (4)				

NOTES: 50% CL estimates provided by PMW 150 for each of the planned releases. Higher SLOC estimates were provided by the MTC2 Software Support Activity. The estimates do not account for or quantify Alternative 4 cost impacts of mitigating the potential schedule risks of NTC fielding delays.

the scope and/or complexity of the S/W development efforts required for generating higher fidelity estimates especially for cloud-based Alternatives 3 and 4.

For example, after reviewing the MTC2 AoA study-plan descriptions of Alternatives 3 and 4 as guidance, at least three major differences become readily apparent in estimating the development effort for these two new MTC2 alternatives. In Table 5.6, we summarize the four cost-discriminator areas and identify potential outcomes that would affect the magnitude of the estimated development or demonstration testing effort and associated cost and/or schedule. For Alternatives 3 and 4, we list the specific summarized findings and the relative magnitudes of increasing (yellow or red arrows up) or decreasing (green arrow down) cost impacts for each.

Overall Risk

MTC2 risks were broken into four categories: technical, schedule, programmatic, and cost. Analysts used an iterative process to determine what these risks should be, then used a second iterative process to score the probability and consequence of these risks in relation to the four MTC2 alternatives under consideration in this report. The remainder of this chapter explores how risk scores for technical, schedule, programmatic, and cost risks were determined.

Alternatives 1 and 2 concluded that they faced little to no schedule, programmatic, or cost risk. This is to be expected, as Alternative 1 is the status quo and Alternative 2 adds C2RPC to the status quo. C2RPC is a tool that is already functional, so it makes sense that no great risk to schedule or program management and cost should come from continuing to use currently available tools. Alternatives 1 and 2, however, face severe technical risk. Their reliance on the current data architecture results in very inefficient movement and management of an ever-increasing amount of data and number of data sources. Collaboration and DIL functionality are also severely limited.

Alternative 3 alleviates some technical risk from Alternatives 1 and 2 with the use of an ashore cloud. However, many of the issues arising from the stove-piped afloat architecture still exist, such as poor DIL performance, redundant data feeds, and multilevel security solutions. Alternative 3 adds risk from the need to integrate the legacy afloat architecture with the new ashore cloud architecture and further programmatic and schedule risks arise from programs and technologies related to functionality and management of the ashore cloud. Alternatives 3 and 4 have greater cost risk than Alternatives 1 and 2, but it should be noted that this risk is only moderate.

Alternative 4 proves to be the least technically risky of the alternatives. Its afloat and ashore clouds enable reliable data access and management as well as functionality in DIL environments. Alternative 4's high risks come from programmatic and schedule concerns. Alternative 4 has many of the same programmatic and schedule risks as Alternative 3 and adds even more of these types of risk with the addition of NTC. Some cloud-related schedule and programmatic risks are also judged to be of greater consequence and, as such, higher risk for Alternative 4 because of its total reliance on the cloud.

As Table 5.7 shows, Alternative 4 has the lowest technical risk, while Alternatives 1 and 2 have the lowest schedule, programmatic, and cost risks. High technical risks are unlikely to change, and many of them will only grow worse as the Navy adds more sensors and data feeds. Seeing system failure with no workarounds due to antiquated data architecture is not outside the realm of possibility if Alternative 1 or 2 is chosen. Alternative 3 faces the same threat to a lesser degree due to keeping a non-cloud afloat architecture. Cost risk is a lesser concern, since no alternative has high cost risk. These reasons are why Alternative 1 is judged to be the most risky of the four alternatives, and Alternative 4 the least risky for MTC2.

5.4. IA Analysis Results

The IA analysis (see Table 5.8) was tasked to consider the relative effort required for each of the MTC2 alternatives to achieve and maintain IA compliance. The evaluation was based on the following five criteria:

- **Security architecture finalization date**. Architectures finalized more than three years ago are likely to require more effort

Table 5.6
Comparisons of Alternative 3 and 4 Qualitative Cost Discriminators and Potential Impacts on Updating Cost Estimates

Cost Discriminators	Potential Outcomes	Alternative 3	Cost Impact	Alternative 4	Cost Impact
Leveraging previously demonstrated technology	Reduced development effort	(1) NSA ghost machines (2) DCGS-A (3) ONR Magic Mirror Demo	↓	None currently apparent	?
Leverage ongoing development activities	(1) Higher system I/F complexity (2) Schedule-Dependency related risk mitigation	Operate afloat and mobile units at Alternative 2 level	↑	(1) ONR tactical cloud; (2) transition-applicable C2RPC capabilities	↑ ↑
Unique capabilities and added effort or interface complexity	Increased development effort	Requires data ingest from current sources used by C2RPC	↑	Requires sizing additional storage at afloat sites	↑
Unique operating environment	Increased demo testing effort	N/A		Ensure operations in DIL and A2/AD	↑

to achieve and maintain compliance. Updates to DoD IA regulations and security technical implementation guides over the past three years have dramatically raised the IA compliance bar.

- **Inherited IA protections**. Programs that properly leverage IA protections from other programs can realize cost savings in development.
- **Security posture maturity**. Alternatives with a mature security posture will likely require less work in order to achieve the required state of IA compliance.
- **Sustainability of security posture**. Alternatives with highly sustainable security postures will be more likely to maintain an

acceptable level of IA compliance at a lower cost or for a longer period of time.

- **Level of coupling between mission applications and support infrastructures**. Loosely coupled applications allow infrastructure components to be upgraded independently in order to maintain IA compliance.

Figure 5.10 highlights the overview of the IA compliance considerations, and Table 5.8 provides discussion of each of the conclusions reached in the IA analysis.

Overall, Alternative 4 will be the easiest to achieve and maintain IA compliance. A complete assessment is not possible without a more mature software architecture.

Table 5.7
Overall Comparison of the Four Alternatives

	Risk Type			
	Technical	Schedule	Programmatic	Cost
Alternative 1: Status quo. GCCS-M Increment 2 modernization will be limited	High	Low	Moderate	Low
Alternative 2: Augment GCCS-M Inc. 2 capabilities with productized C2RPC capabilities and newly developed capabilities following C2RPC style and approach	Moderate/ high	Low	Moderate	Moderate/ low
Alternative 3: New system to satisfy maritime C2 requirements as defined in the initial MTC2 guidance. Builds upon the IC data analytic cloud capability. This system will replace GCCS-M at every program of record site	Moderate/ high	Moderate	Moderate/ high	Moderate
Alternative 4: New system to satisfy maritime C2 requirements as defined in the initial MTC2 guidance. Adds the concept of ONR tactical cloud	Moderate	Moderate/ high	Moderate/ high	Moderate

Table 5.8
Discussion of IA Assessment

Criteria	Alternative 1	Alternative 2	Alternative 3	Alternative 4
Security architecture finalization date	Limited new development; however, the measure of risk is how well security controls are imposed at each layer of the system architecture, rather than when the architecture was finalized	Some new development to productize C2RPC; however, the measure of risk is how well security controls are imposed at each layer of the system architecture, rather than when the architecture was finalized	New architecture development; however, the measure of risk is how well security controls are imposed at each layer of the system architecture, rather than when the architecture was finalized	New architecture development; however, the measure of risk is how well security controls are imposed at each layer of the system architecture, rather than when the architecture was finalized
Inherited IA protections	Provides limited opportunity for IA control inheritance	Provides limited opportunity for IA control inheritance	Could provide substantial opportunity for IA control inheritance; however, cloud provider solution undetermined at this time	Could provide substantial opportunity for IA control inheritance. Additional IA controls potentially inherited from CANES (a known entity) for disconnected afloat operations
Security posture maturity	Limited new development; security posture difficult and expensive to maintain	Some new development to productize C2RPC development; need to reassess security posture (Transition Readiness Assessment found C2RPC IA posture lacking)	Need to evaluate how the architecture supports the required security posture, which is unknown at this time and requires further study	Need to evaluate how the architecture supports the required security posture, which is unknown at this time and requires further study

Table 5.8—Continued

Criteria	Alternative 1	Alternative 2	Alternative 3	Alternative 4
Sustainability of security posture	Lack of homogeneity and uniformity of legacy systems increase cost of security maintainability	Lack of homogeneity and uniformity of legacy systems increase cost of security maintainability	Cloud-computing processes offer the potential for improved cyber security; e.g., better traffic filtering and malware scanning, monitoring of usage patterns and end-device configurations, varying provisioning of data resources, and improved management of systems operations (DoD, 2013). Cloud-provider security posture unknown at this time	Cloud computing processes offer the potential for improved cyber security; e.g., better traffic filtering and malware scanning, monitoring of usage patterns and end-device configurations, varying provisioning of data resources, and improved management of systems operations (DoD, 2013). Cloud provider security posture unknown at this time
Level of coupling between mission applications and supporting infrastructures	Tighter coupling increases cost of development and testing required for security maintainability	Tighter coupling increases cost of development and testing required for security maintainability	Cloud-computing architecture provides loose coupling of application and infrastructure components; therefore, components can be upgraded independently in order to maintain IA compliance	Cloud computing architecture provides loose coupling of application and infrastructure components; therefore, components can be upgraded independently in order to maintain IA compliance

Figure 5.10
Summary of IA Compliance Level of Effort Assessment

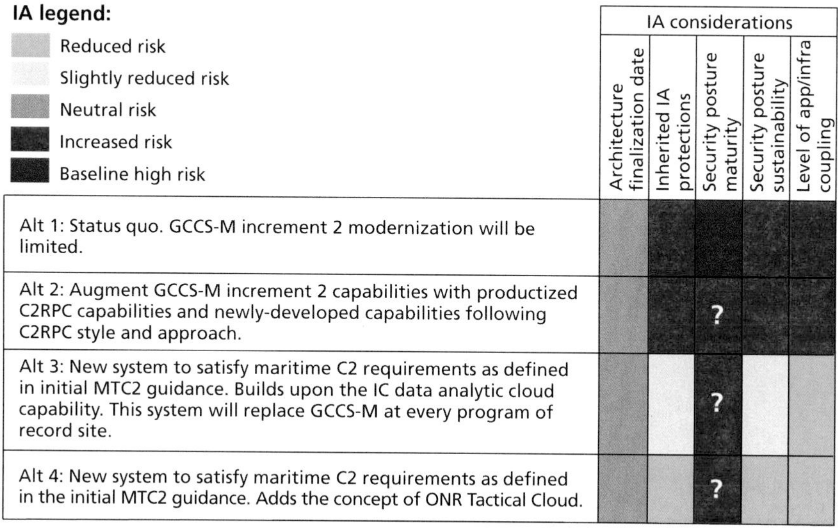

IA legend:	IA considerations				
▨ Reduced risk	Architecture finalization date	Inherited IA protections	Security posture maturity	Security posture sustainability	Level of app/infra coupling
░ Slightly reduced risk					
▩ Neutral risk					
▓ Increased risk					
■ Baseline high risk					
Alt 1: Status quo. GCCS-M increment 2 modernization will be limited.					
Alt 2: Augment GCCS-M increment 2 capabilities with productized C2RPC capabilities and newly-developed capabilities following C2RPC style and approach.			?		
Alt 3: New system to satisfy maritime C2 requirements as defined in initial MTC2 guidance. Builds upon the IC data analytic cloud capability. This system will replace GCCS-M at every program of record site.			?		
Alt 4: New system to satisfy maritime C2 requirements as defined in the initial MTC2 guidance. Adds the concept of ONR Tactical Cloud.			?		

NOTE: "?" indicates no information on IA certification efforts, thus by default making the element high risk.

RAND RR1383-5.10

CHAPTER SIX

Recommended Alternative and Rationale

The results from the performance,[1] cost, and risk analysis are a preference for Alternative 4. It provides pervasive access to information and analytics in a common, globally managed enterprise and demonstrates the greatest potential for achieving C2 workflow efficiencies. The personnel-recovery mission analysis is a good example of Alternative 4's potential because it touches on three of the top four MTs in some capacity: plan collaboratively, develop/maintain shared SA and understanding, and communicate Commander's intent and guidance. The results, though abstracted in this report for the purposes of distribution, demonstrate significant potential gains in Alternative 4 over all other options. Figure 6.1 depicts this.

Alternatives 3 and 4, as new systems for replacing GCCS-M, result in potential annual sustainment cost savings through FY 2030. There is uncertainty in the other cost estimates at the 80-percent confidence level due to uncertainty in the SLOC estimates and other factors (e.g., schedule risk).

IA assessments acquired through in-person interviews suggests that Alternatives 3 and 4 will be relatively easier to achieve and maintain IA compliance than Alternatives 1 and 2, but a more-thorough IA analysis is needed. Figure 6.2 summarizes the results.

Maintaining the status quo supposes the least amount of risk in cost and schedule, but at a monetary cost that is higher than it needs to be given the Navy's migration to CANES and a performance cost

---

[1] Including modeling and simulation, static analysis, and the tabletop exercise.

Figure 6.1
Performance of the Develop Personnel Recovery Mission Workflow

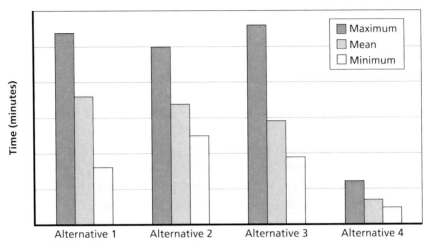

NOTE: Axis values have been abstracted.
RAND RR1383-6.1

whereby GCCS-M cannot meet a number of the critical performance measures identified by the Navy. Alternative 4 carries some risk, as do most of the alternatives in at least one area. Alternative 4 may have high consequence in delayed fielding of the NTC or more-capable data-integration systems, but this is a given for a future, less stove-piped Navy. Reducing overall cost through a shared architecture will increase programmatic and schedule complexity, but, if managed well, this report shows that there is a potential future where the Navy is able to reduce overall life-cycle costs and achieve significant performance gains. Figure 6.3 highlights the key risks for all alternatives.

Given this analysis, and given that the Navy is already embarking on providing shared infrastructure to support other programs that MTC2 can leverage, we recommend Alternative 4.

Figure 6.2
AoA Summary Results

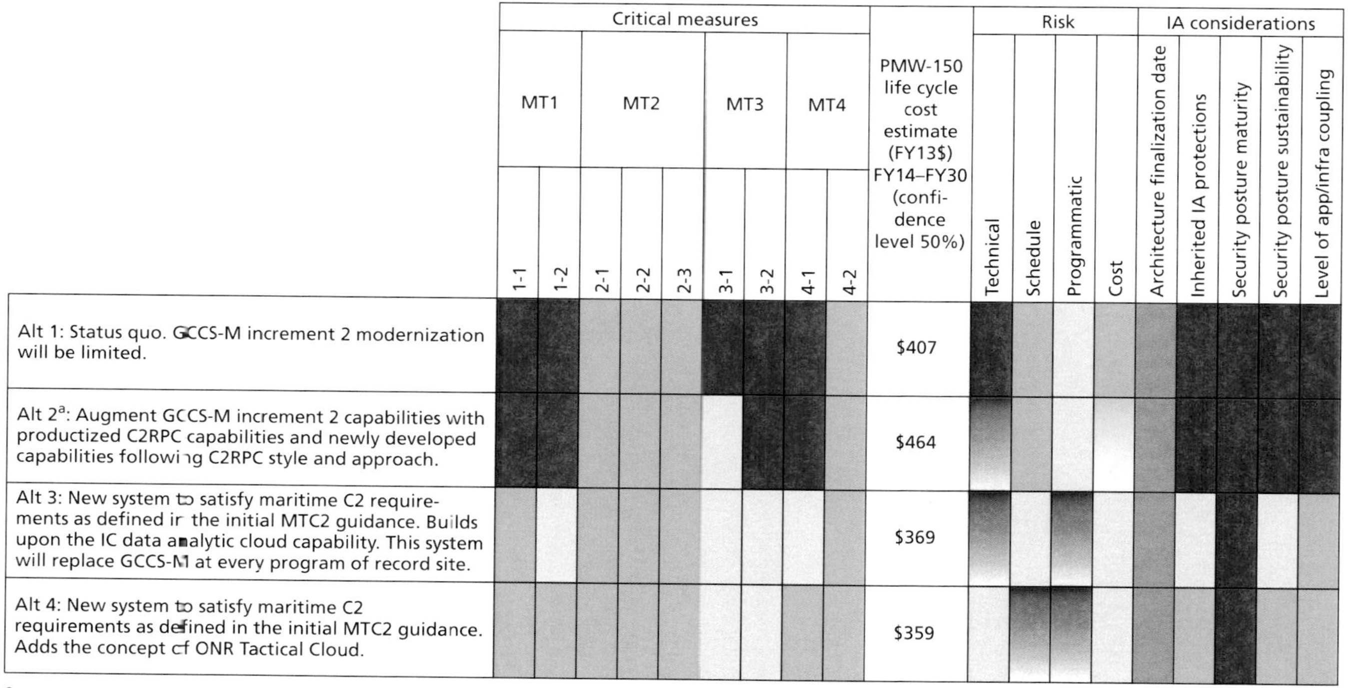

	Critical measures									PMW-150 life cycle cost estimate (FY13$) FY14–FY30 (confidence level 50%)	Risk				IA considerations				
	MT1		MT2			MT3		MT4			Technical	Schedule	Programmatic	Cost	Architecture finalization date	Inherited IA protections	Security posture maturity	Security posture sustainability	Level of app/infra coupling
	1-1	1-2	2-1	2-2	2-3	3-1	3-2	4-1	4-2										
Alt 1: Status quo. GCCS-M increment 2 modernization will be limited.										$407									
Alt 2[a]: Augment GCCS-M increment 2 capabilities with productized C2RPC capabilities and newly developed capabilities following C2RPC style and approach.										$464									
Alt 3: New system to satisfy maritime C2 requirements as defined in the initial MTC2 guidance. Builds upon the IC data analytic cloud capability. This system will replace GCCS-M at every program of record site.										$369									
Alt 4: New system to satisfy maritime C2 requirements as defined in the initial MTC2 guidance. Adds the concept of ONR Tactical Cloud.										$359									

[a] For this report, Alternative 2 assumes newly developed capabilities, "2+."

RAND RR1383-6.2

Figure 6.3
All Alternatives Have Some Risk

Alternative 1 has significant risk from

- *Technical*: denied communication environments
- *Technical*: need for a multilevel security solution
- *Technical*: need for collaborative planning tools
- *Technical*: ability to deal with increasing number of data sources
- *Technical*: need for automation technology
- *Technical*: need for reliable, redundant data feeds
- *Programmatic*: need for an effective data strategy

Alternative 2 has significant risk from

- *Technical*: denied communication environments
- *Technical*: need for a multilevel security solution
- *Technical*: need for automation technology
- *Technical*: need for reliable, redundant data feeds
- *Technical*: ability to deal with increasing number of data sources
- *Technical*: need for collaborative planning tools
- *Programmatic*: need for an effective data strategy

Alternative 3 has significant risk from

- *Technical*: denied communication environments
- *Schedule/programmatic*: need for a capable data integration layer
- *Programmatic*: need for an effective data strategy
- *Technical*: need for a multilevel security solution
- *Programmatic*: need for a cloud architecture governance strategy
- *Technical*: need for automation technology
- *Technical*: need for reliable, redundant data feeds
- *Programmatic*: need for user-defined queries/information management (DQS)
- *Schedule*: delay fielding of critical technologies services (IDAM, ABAC, ACS)

Alternative 4 has significant risk from

- *Schedule*: delay fielding of Naval Tactical Cloud
- *Schedule/programmatic*: need for a capable data integration layer
- *Programmatic*: need for an effective data strategy
- *Programmatic*: need for a cloud architecture governance strategy
- *Technical*: need for automation technology
- *Programmatic*: need for user-defined queries/information management (DQS)
- *Technical*: need for a multilevel security solution
- *Schedule*: delay fielding of critical technologies services (IDAM, ABAC, ACS)

APPENDIX

Tabletop Exercise Overview

Background

The tabletop exercise described in this appendix was held in the RAND Corporation's Washington office on May 9, 2013. Personnel from N2/N6, ONR and RAND were involved. In advance of the meeting, RAND provided handouts, read-aheads via email, and conducted teleconferences. The results of the exercise did not differ from the other analyses. The conclusion is that Alternative 4 is the preferred alternate, especially given the consideration of A2AD.

Problem Overview

A key purpose of an AoA is to identify discriminators between material solution alternatives for the MDA.

Typical AoA analyses include:

- performance analysis (typically modeling) to identify likely technical/operational performance differences between alternatives
- cost analysis to assess the likely life-cycle cost differences between alternatives
- risk assessment to identify key elements of uncertainty that could impact the program's ability to meet engineering, cost, performance, and/or schedule objectives.

Challenge

Understanding how key implementation differences could impact application functionality and operational performance from the perspective of operators, products, and decisionmakers.

Tabletop Exercise Objectives and Approach

Objectives of the tabletop exercise were to:

- Validate the most-important differences between alternatives that could impact application functionality and operational performance from the operator, C2 product, and decisionmaker perspectives.
- Assess potential performance differences between alternatives in the context of the government-approved scenario, against government-approved MoEs, and/or MoPs.

The approach is to facilitate a discussion with government-selected operational and technical SMEs. It included:

- Review and refine key implementation differences between alternatives that could impact application functionality and operational performance from the operator, C2 product, and decisionmaker perspectives.
- Select those implementation factors that could highlight performance differences as expressed in one or more MoEs.
- Discuss the potential performance impacts and causes.
- Assess the likely relative performance differences between alternatives.

Context

Scenario

- 2016 MSFD

Vignettes

- Representative vignettes were chosen that would be expected to stress one or more JIC metrics.

Tabletop Exercise Metrics Definitions
Accessibility
The ability of all levels of command (strategic, operational, and tactical), at any time and from anywhere, to pull or push relevant data and information that is the basis for shared SA. Additionally, access to a standardized joint application tool set at austere and robust, fixed and mobile sites will enhance decisionmaking capabilities supporting rapid, efficient, effective C2.

Accuracy
Conforming precisely to fact or truth. A system with this attribute provides error-free (or within a range of acceptable error) measurements or data via credible, dependable, and reliable sources. Accuracy and trust may exist due to prior performance and/or specific integrity-assurance measures that have been adopted.

Completeness
Having all components, parts, or steps critical to complete an operation. Complete information enables timely and appropriate decisionmaking.

Interoperability
The ability of systems, units, or forces to provide data, information, materiel, and services to and accept the same from other systems, units, or forces. Additionally data, information, materiel, and services can be used to enable them to operate effectively together. Information technology and National Security System interoperability also includes

both the technical exchange of information as required for mission accomplishment.

Responsiveness

Readily reacting to or recovering from changing situations and conditions in real time and near real time. The effective use of responsive and resilient planning, execution, and assessment enables rapid deployment or redirection of assets when various "windows of opportunity" occur. Ideally, systems with this attribute are designed to function at their normal operational standard upon recovery from or reaction to changing situations and conditions.

Speed

The appropriate pace of tasks and decisionmaking. At times, the appropriate speed is rapid. When deliberate methodical actions are required, a slower speed may be required. To obtain the appropriate speed of command, subordinate forces must be enabled to synchronize actions among themselves without restrictive direction from above.

Timeliness

Occurring at a suitable or opportune moment; well timed. Timeliness is situation dependent. It reflects the relationship between the age of an information item and the tasks or missions it must support (DoD, 2005) This subset represents metrics associated with MoEs-MoPs approved for AoA analysis.

Table A.1 helps to discriminate key implementation factors between the alternatives for the participants in the tabletop exercise. Each dot indicates which alternative the implementation factor maps to.

Table A.1
Key Implementation Factors

Implementation Factors	Alt. 1	Alt. 2	Alt. 3	Alt. 4
Functional/application architecture: How application functionality will be implemented				
Modernization limited to bug fixes	●			
Productized C2RPC capabilities and newly developed capabilities following C2RPC style and approach		●		
New software using agile development techniques with continuous end-user involvement			●	●
Continuous modernization			●	●
Data strategy/architecture: What, how, and where data are ingested, processed, stored, and made accessible to C2 applications				
All C2 nodes retain GCCS-M legacy data model (including database architecture; Extract, Transform, Load [ETL], and point-to-point service-level agreements [SLAs])	●	●		
Afloat and mobile nodes retain GCCS-M legacy data model (including database architecture, ETL, and point-to-point SLAs)			●	
Shore C2 nodes leverage IC Data Analytic Cloud (federated data discovery system [FDDS] and automated Extract, Load, Transform [ELT])			●	●
Afloat nodes leverage NTC (automated ELT; ingest of ships organic C2 data; shore-sync; and Distributed Query System [DQS])				●
IA strategy (including identity and access management): What data C2 users and applications are permitted to use				
Multiple security enclaves		●	●	●
IDAM and ABAC			●	●
Continuous data validity reassessment			●	●

Mission Task to Metrics to MoEs-MoPs

Table A.2 lists the eight MT areas and their mapping to an MoE. The first four were isolated in the analysis for this report; however, this appendix expounds on the bottom for tasks as they were considered in the tabletop exercise.

Tables A.3 through A.6 discuss the source and assumption about each alternative for the exercise participants. It further documents what the change was in comparing the alternatives.

Tabletop Exercise Discussion Guidance and Assumptions

Operational Context

Operational examples were constrained to the scenario and associated vignettes.

MTC2 Functional Capabilities

Assume the MTC2 design-development priorities are weighted as follows:

1. Plan collaboratively.
2. Develop and maintain shared SA and understanding.
3. Establish/adapt command structures and enable both global and regional collaboration.
4. Communicate Commander's intent and guidance.
5. Exercise command leadership.
6. Synchronize execution across all domains.
7. Monitor execution, assess effects, and adapt operations.
8. Leverage mission partners.

Discussion Guidance/Focus

- How can key implementation factors enable or prevent implementation of certain functionality?

Table A.2
Mission Tasks to Metrics to MoE

Plan collaboratively	Interoperability	Percentage of data and information that is interoperable between JIIM partners	Percentage of five-eye partner C2 data/information that is interoperable
			Percentage of DoD C2 data/information that is interoperable
			Percentage of U.S. interagency C2 data/information that is interoperable
	Timeliness	Timeliness of planning information dissemination to JIIM	Time for planning information dissemination to interagency
			Time for planning information dissemination between joint partners
			Time for planning info dissemination to multinational partners
Develop/maintain shared SA and understanding	Completeness	Completeness of COP (percentage of forces)	Number of blue forces on COP (plotted over time)
			Number of white ships on COP (plotted over time)
	Accuracy	Percentage of friendly force locations that are accurate	Blue force location error (average)
			Neutral force location error (average)
	Bandwidth efficiency	Percentage of available bandwidth consumed	Data rate over links during mission

Table A.2—Continued

Establish/adapt command structures and enable both global and regional collaboration	Interoperability	Percentage of C2 data interoperability across tactical and operational echelons	Five-eye data interoperability
			DoD data interoperability
			Multinational data interoperability
			U.S. interagency data interoperability
	Responsiveness	Likelihood of reconfigurability to dynamic mission requirements and return to steady state	Percentage of time mission partners can reconfigure in response to dynamic mission requirements
Communicate commander's intent and guidance	Timeliness	Percentage of commander's plans received by appropriate maritime personnel	Timeliness of promulgation of guidance
		Percentage of time orders are received in time to conduct the task/mission	Timeliness of promulgation of guidance
	Speed	Percentage of time orders are received in time to conduct the task/mission	Time to conduct the task/mission

Table A.2—Continued

Exercise command leadership	Timeliness	Number of readiness assessments completed in time	Volume of readiness assessments
	Accuracy	Percentage of released information that is correct (accuracy)	Volume of released information
			Volume of released information with inaccuracies
	Speed	Time to promulgate RoE and rules for the use of force changes	Five-eye data promulgation time
			Multinational data promulgation time
			U.S. interagency data promulgation time
Synchronize execution across all domains	Accessibility	Percentage of subordinate forces able to access unclassified information at operational level (accessibility)	Time required to get access to unclassified information
		Percentage of subordinate forces able to access unclassified information at the tactical level (accessibility)	Time required to get access to unclassified information
	Timeliness	Plans are completed, disseminated, received in time	Percentage availability of C2 node health to maritime commanders

Table A.2—Continued

Monitor execution, assess effects, and adapt operations	Completeness	Number of fires processes, networks, and systems MOC can efficiently track	Percentage of assets tracked (completeness)
	Speed	Percentage of forces and assets that can quickly change operations to facilitate direction change (agility)	Time it takes to communicate a needed change Time limit on when operations have to be modified
	Timeliness	Percentage of forces and assets that can quickly change operations to facilitate direction change (agility)	Number of units directed to change Time it takes to modify operations by units upon direction
Leverage mission partners	Interoperability	Percentage of JIIM partners MOC can exchange information with	Percentage of five-eye partner C2 data/information that is interoperable Percentage of DoD C2 data/information that is interoperable Percentage of multinational partner C2 data/info that is interoperable Percentage of U.S. interagency C2 data/information that is interoperable
		Percentage of mission partners that receive and understand commanders intent	Percentage of personnel that receive guidance

SOURCES: Initial MTC2 guidance; DoD, 2005; and "PMW 150 Command and Control Systems Program Office," 2015.

Table A.3
Alternative 1 Characteristics: Status Quo GCCS-M Increment 2 (Baseline)

Assumptions and Considerations	Source	Specified/Implied
GCCS-M Increment 2 projected costs	AoA guidance	Specified
GCCS-M Increment 2 projected capabilities plus programmed enhancements	AoA guidance	Specified
No follow-on capability developments into GCCS-M Increment 2	AoA guidance	Specified
No migration or integration of additional C2 solutions into GCCS-M Increment 2	AoA guidance	Specified
Modernization limited to bug fixes, IA-related fixes, and patches and alignment with COTS/GOTS hardware and software technology refresh	AoA guidance	Specified

SOURCE: Tighe, 2012.

- What can affect MTC2 functionality and/or performance; how and where data are ingested, processed, stored, and made accessible?
- How MTC2 functionality and/or performance could be affected by enterprise IA strategy that would determine/constrain what data C2 users and/or applications are permitted to use?

Table A.4
Alternative 2 Characteristics: GCCS-M Increment 2 Augmented with C2RPC-Like Capabilities

Assumptions and Considerations	Source	Specified/ Implied	Delta from Previous Alternative
GCCS-M Increment 2 projected costs	AoA guidance	Implied	
GCCS-M Increment 2 projected capabilities plus programmed enhancements	AoA guidance	Implied	
Augment GCCS-M Increment 2 SA capabilities with productized C2RPC capabilities and newly developed capabilities following C2RPC style and approach	AoA guidance	Specified	Add
Provide a much wider range of C2 capabilities from MOCs down to tactical level	AoA guidance	Specified	Add
New software will be installed and integrated with GCCS-M on afloat and ashore GCCS-M infrastructures	AoA guidance	Specified	Add
Modernization will be limited to bug fixes, IA-related fixes, and patches and alignment with COTS/GOTS hardware and software technology refresh	AoA guidance	Specified	

SOURCE: Tighe, 2012.
NOTE: Green text indicates additions from the baseline.

Tabletop Exercise Results

Tables A.7 and A.8 detail the results for the critical measures. The color-coding thresholds are at the bottom of each table.

Table A.5
Alternative 3 Characteristics: New System Development with Ashore Data Analytic Cloud

Assumptions and Considerations	Source	Specified/Implied	Delta from Previous Alternative
Satisfy JIC and initial MTC2 guidance for maritime C2 requirements from operational level down to tactical edge	AoA guidance	Specified	Change
Maintain backward compatibility with existing GCCS-M systems	AoA guidance	Specified	Change
Transition applicable C2RPC capabilities	AoA guidance	Specified	Change
Builds on IC data analytic cloud	AoA guidance	Specified	Change
Data analytic cloud will continuously ingest data sources used by C2RPC	AoA guidance	Specified	Change
Eliminate need for each C2 node to independently search and integrate large numbers of databases and web pages	AoA guidance	Specified	Change
Data analytic cloud will act as a clearing house, automatically and continuously collects, processes, and stores data in response to user actions	AoA guidance	Specified	Change
Afloat and mobile nodes will continue to operate at Alternative 2 level	AoA guidance	Specified	Change
Shore node will act as preprocessing capability to tailor and enhance operational capabilities	AoA guidance	Specified	Change
System will replace GCCS-M at every POR site without loss of legacy operational functionality and capability	AoA guidance	Specified	Change

Table A.5—Continued

Assumptions and Considerations	Source	Specified/Implied	Delta from Previous Alternative
New software development will follow agile development techniques, with continuous end-user involvement and responsiveness to fleet requirements	AoA guidance	Specified	Change
Continuous modernization will include bug fixes, IA-related fixes, and patches and alignment with COTS/GOTS hardware and software technology refresh	AoA guidance	Specified	Change

SOURCE: Tighe, 2012.

NOTE: Blue text indicates changes from the baseline.

Table A.6
Alternative 4 Characteristics: New System Development with Naval Tactical Cloud

Assumptions and Considerations	Source	Specified/Implied	Delta from Previous Alternative
Satisfy JIC and initial MTC2 guidance for maritime C2 requirements from operational level down to tactical edge	AoA guidance	Implied	
Maintain backward compatibility with existing GCCS-M systems	AoA guidance	Implied	
Transition applicable C2RPC capabilities	AoA guidance	Implied	
Builds on IC data analytic cloud	AoA guidance	Implied	
Data analytic cloud will continuously ingest data sources used by C2RPC	AoA guidance	Implied	
Eliminate need for each C2 node to independently search and integrate large numbers of databases and web pages	AoA guidance	Implied	
Data analytic cloud will act as a clearing house, automatically and continuously collects, processes, and stores data in response to user actions	AoA guidance	Implied	
(Removes) Afloat and mobile nodes will continue to operate at Alternative 2 level	AoA guidance	Specified	Remove
Adds data analytic cloud afloat (NTC)	AoA guidance	Specified	Add
Provides significant additional storage space for afloat units, preloaded with historic information	AoA guidance	Specified	Add

Table A.6—Continued

Assumptions and Considerations	Source	Specified/Implied	Delta from Previous Alternative
Tactical cloud will allow for continuous synchronization between afloat tactical nodes and shore data analytic nodes	AoA guidance	Specified	Add
Intended to ensure operations in A2AD environments	AoA guidance	Specified	Add
Includes automated ingestion of ships organic C2 data	AoA guidance	Specified	Add
Ingested organic C2 data can be synchronized with shore nodes for a more complete and timely SA picture	AoA guidance	Specified	Add
System will replace GCCS-M at every POR site without loss of legacy operational functionality and capability	AoA guidance	Specified	
New software development will follow agile development techniques, with continuous end-user involvement and responsiveness to Fleet requirements	AoA guidance	Specified	
Continuous modernization will include bug fixes, IA-related fixes and patches, and alignment with COTS/GOTS hardware and software technology refresh	AoA guidance	Specified	

SOURCE: Tighe, 2012.

NOTE: Green text indicates additions from the baseline.

Table A.7

TTX Results: Normal Environment—Critical Measures

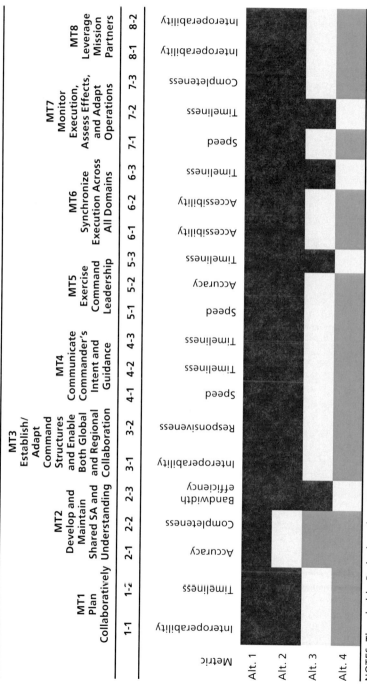

NOTES: Threshold: Red = less than 25-percent projected improvement over Alternative 1; yellow= 25–44-percent projected improvement over Alternative 1; green = more than 45-percent projected improvement over Alternative 1.

Table A.8
TTX Results: DDIL Environment—Critical Measures

Metric	MT1 Plan Collaboratively		MT2 Develop and Maintain Shared SA and Understanding			MT3 Establish/Adapt Command Structures and Enable Both Global and Regional Collaboration		MT4 Communicate Commander's Intent and Guidance			MT5 Exercise Command Leadership			MT6 Synchronize Execution Across All Domains			MT7 Monitor Execution, Assess Effects, and Adapt Operations			MT8 Leverage Mission Partners	
	1-1 Interoperability	1-2 Timeliness	2-1 Accuracy	2-2 Completeness	2-3 Bandwidth Efficiency	3-1 Interoperability	3-2 Responsiveness	4-1 Speed	4-2 Timeliness	4-3 Timeliness	5-1 Speed	5-2 Accuracy	5-3 Timeliness	6-1 Accessibility	6-2 Accessibility	6-3 Timeliness	7-1 Speed	7-2 Timeliness	7-3 Completeness	8-1 Interoperability	8-2 Interoperability
Alt. 1																					
Alt. 2																					
Alt. 3																					
Alt. 4																					

NOTES: Thresholds: Red = less than 25-percent projected improvement over Alternative 1; yellow = 25–44-percent projected improvement over Alternative 1; green = more than 45-percentage projected improvement over Alternative 1.

References

Akins, Diana, Program Executive Office, Command, Control, Communications, Computers and Intelligence (PEO C4I), "C2 Rapid Prototyping Continuum (C2RPC)," Microsoft PowerPoint presentation provided to authors, April 19, 2011.

Defense Information Systems Agency (DISA), *Fiscal Year 2011 Budget Estimates*, February 2010. As of April 14, 2016:
http://comptroller.defense.gov/Portals/45/Documents/defbudget/fy2011/budget_justification/pdfs/01_Operation_and_Maintenance/O_M_VOL_1_PARTS/DISA_FY11.pdf

DoD—*See* U.S. Department of Defense.

Gartner, "Gartner IT Glossary: Identity and Access Management (IAM), undated. As of April 14, 2016:
http://www.gartner.com/it-glossary/identity-and-access-management-iam/

"GCCS-M_PLCCE_17September2012_iCRB_with 1-K risk run," Microsoft Excel document provided to authors by the U.S. Navy, September 17, 2012.

Joint Chiefs of Staff, *Joint Publication 3-33, Joint Task Force Headquarters*, July 30, 2012. As of April 14, 2016:
http://www.dtic.mil/doctrine/new_pubs/jp3_33.pdf

Joint Staff, *Joint Command and Control (C2) Requirements Management Process and Procedures*, Washington, D.C., CJCSM 3265.01A, November 29, 2013.

"Maritime Tactical Command and Control (MTC2): Analysis of Alternatives (AoA) Study Plan," facsimile provided to authors by the U.S. Navy, January 15, 2013, approved on February 7, 2013.

McDonnell, John, "C2RPC Transition Readiness Assessment (XNRA) Findings and Recommendations, MTC2 Software Support Activity," Microsoft PowerPoint briefing slides provided to authors, March 14, 2014.

National Institute of Standards and Technology, "Attribute Based Access Control Overview," May 6, 2015. As of May 10, 2016:
http://csrc.nist.gov/projects/abac

Office of Aerospace Studies, *Analysis of Alternatives (AoA) Handbook: A Practical Guide to Analyses of Alternatives*, Kirtland Air Force Base, N.M.: Air Force Materiel Command (AFMC) OAS/A9, July 2008.

———, *Analysis of Alternatives (AoA) Handbook: A Practical Guide to Analyses of Alternatives*, Kirtland Air Force Base, N.M.: Air Force Materiel Command (AFMC) OAS/A9, July 2010. As of April 14, 2016: http://www.prim.osd.mil/Documents/AoA_Handbook.pdf

Office of Naval Research, "ONR Develops New Acquisition Model for Delivering Information to the Fleet," Arlington, Va., February 10, 2011. As of April 20, 2016: http://www.onr.navy.mil/en/Media-Center/Press-Releases/2011/C2RPC-Information-Fleet.aspx

———, "Command and Control Rapid Prototyping Capability," 2012. As of April 14, 2016: http://www.onr.navy.mil/Media-Center/Fact-Sheets/C2RPC.aspx

ONR—*See* Office of Naval Research.

"OPNAV Study Guidance," January 2013.

"PMW 150 Command and Control Systems Program Office," January 2015. As of April 14, 2016: http://www.public.navy.mil/spawar/PEOC4I/Documents/Tear%20Sheets/PMW%20150_Tear%20Sheet_JAN2015-approved.pdf

Tighe, Jan, "Maritime Tactical Command and Control (MTC2) Analysis of Alternatives (AoA) Study Guidance," Action Memo, facsimile provided to authors, November 19, 2012.

U.S. Department of Defense, *Command and Control, Joint Integrating Concept*, Final Version 1.0, September 1, 2005. As of June 27, 2016: http://www.dtic.mil/futurejointwarfare/concepts/c2_jic.pdf

U.S. Department of Defense, Department of Defense Science Board, Office of the Under Secretary of Defense for Acquisition, Technology, and Logistics, *Task Force Report: Cyber Security and Reliability in a Digital Cloud*, Washington, D.C., January 2013. As of April 20, 2016: http://www.acq.osd.mil/dsb/reports/CyberCloud.pdf

Walsh, P. J., J. W. Bailey, R. A. Keller, C. H. Lyman, K. A. Morrison, R. B. Polin, G. A. Sharp, J. R. Shea, D. Spalding, and R. W. Carpenter, *Joint Command and Control (JC2) Capability Analysis of Alternatives (AoA)*, Volume I—Main Report and Two Appendixes, Alexandria, Va.: Institute for Defense Analyses, IDA Paper P-3996, May 2005, not available to the general public.

Wellman, John, "Joint Command and Control in a Net-Centric Environment," Microsoft PowerPoint slides, 2005. As of July 31, 2016: http://www.dtic.mil/ndia/2005netcentric/thursday/wellman.pdf